Users' Guides to H
and Ergonomics Methods

Usability Assessment:

How to Measure the Usability of Products, Services, and Systems

Philip Kortum

PUBLISHED BY THE HUMAN FACTORS AND ERGONOMICS SOCIETY

To Rebecca

ISBN 978-0-945289-49-4

Published by the
Human Factors and Ergonomics Society
P.O. Box 1369, Santa Monica, CA 90406-1369 USA
http://hfes.org

The Human Factors and Ergonomics Society is a multidisciplinary professional association of more than 4,500 persons in the United States and throughout the world. Its members include psychologists, designers, and scientists, all of whom have a common interest in designing systems and equipment to be safe and effective for the people who operate and maintain them.

Library of Congress Cataloging in Publication Data
On file with publisher

Users' Guides to Human Factors and Ergonomics Methods

TABLE OF CONTENTS

FOREWORD

I am excited that the Human Factors and Ergonomics Society is supporting the development of a book series intended to help students, practitioners, and researchers develop their arsenal of methods. Some readers will use the *Users' Guides to Human Factors and Ergonomics Methods* to improve a basic understanding; other readers may find themselves being exposed to a methodology that they have never used. My hope is to begin a series of guides that make the methods of human factors engineering and engineering psychology broadly accessible.

The methodology of a discipline is a central characteristic both in defining that discipline and in distinguishing it from others. The subject matter of a discipline alone is not sufficient to uniquely identify it. If someone studies human behavior, is she a psychologist or an anthropologist? If someone studies chemical reactions, is he a chemist or a petroleum engineer? If I study the cognitive processes of people while they work with computers, am I an engineering psychologist or a cognitive psychologist?

For much of my career, I considered myself an applied cognitive psychologist. My team and I (all cognitive psychologists) worked extensively on air traffic control, addressing applied questions from the Federal Aviation Administration. We were aware that much work was done by engineering psychologists and human factors professionals, but our approach was quite different. We had cognitive psychology goals and used cognitive psychology methods.

However, we were moving outside the lab and using engaged operators instead of uninterested undergraduates. Applying cognitive psychology in the uncontrolled world of interested people was exciting. This applied cognitive work was different from the work we did in the lab, but it was also different from the work done by engineering psychology colleagues and human factors engineers who also studied aviation and air traffic control. We began to more frequently borrow methods rarely used by applied cognitivists (e.g., hierarchical task analysis, link analysis). This brought more insights, requiring additional human factors methods (e.g., risk analysis, use of standards, qualitative methods).

Our team was well versed in the domain of air traffic control, understood its problems, appreciated others' research perspectives, and was able to contribute to the research needs of the FAA. Yet, until I had mastered the methodology of human factors, I felt as if I were pretending to do applied work.

A conversation with Doug Herrmann helped me appreciate this angst. Herrmann had highlighted the difference between applicable research and applied research. As an applied cognitive psychologist, I certainly was eager to help the FAA, and I felt comfortable finding a path that might lead to a solution; well, frankly, might lead someone else - someday - to a solution. And the methods I used at that time did not allow me to do more. Thus, as an applied cognitive psychologist, I was comfortable doing applicable research.

Now, there is nothing wrong with applicable research, as Seinfeld might say, and I continue to do that type of research and encourage my students to appreciate the value of theoretical work that has broad generality. We continue to do work in what Stokes called Pasteur's quadrant - "use inspired basic research." Indeed, all our PhD students at Georgia Tech must produce a dissertation that contributes to both theory and practice.

Once human factors/ergonomics (HF/E) methods had been added to our arsenal, we were able to do applied, not just applicable, research. The goal (we do not always succeed) of the research is to find a solution that can be implemented, and implemented soon; perhaps some additional work would be needed, but the path to application should be visible. In short, our ability to produce applied research depends on HF/E methods.

HF/E methods include both quantitative and qualitative procedures. A researcher might collect precise response times, whereas another might analyze the opinions and insights from interviews. One might focus on determining causality, emphasizing internal validity. Another might focus on external validity. Yet a third might forego external validity, forced to leverage the data from a low-*N* procedure, and provide important insights into rare events experienced by rare people.

Genesis of the Series

The proposal that led to the volume in your hands was made at the Executive Council of the Human Factors and Ergonomics Society in 2013 during the part of the two-day meeting I which the president-elect offers new projects and new directions for the Society. I had always been a fan of the SAGE "green books" on quantitative methods and thought that the methods we use in HF/E would also lend themselves to a similar approach.

I remember, as a new faculty member, using a SAGE green book to move from a mere awareness of the existence of multidimensional scaling to an understanding of its procedures and details of related analyses as well as an appreciation of the subtle cues in the problem and the nuanced decisions that needed to be made along the way. I went on to publish and present my own MDS research based almost entirely on that little green book. I wanted to offer that experience for the methodologies that characterize HF/E.

The Executive Council enthusiastically supported the idea of developing a series of methods volumes - they actually applauded, as I recall. (I would discover in my year as president of the Society that this would prove atypical.) Council agreed there was a need for the series. Even though there were several methods offerings out there, it remained difficult to find a source that would enable the reader to conduct a study using the method. We discussed those professionals in the workforce who find themselves in a position in which they are expected to perform a human factors investigation or audit that they had never been trained to do. We discussed students who only heard about a particular method or maybe read a two-page description of it, but who, at the end of that reading, did not really know how to execute the procedure.

But truthfully, I think the enthusiasm for the series came from our own wish lists. For me, I have always wanted to know - really know - how to do a risk analysis. I have certainly read about them, but I still did not have enough understanding of the nuances and subtleties to actually conduct one. I envisioned a volume in this series finally giving me the clarity to proceed. Perhaps each Council member entertained a similar desire.

Indeed, there are nuances and subtleties guiding decisions in methodologies that make a difference between a successful use of a method and an unsuccessful one. I wanted to create a series of users' guides that put an expert in readers' heads, giving them the advantage of the expert's experience, insights into how to mitigate threats to the research mission, how to traverse bumps in the procedural details....in other words, I wanted an adviser over my shoulder—yet another voice in my head.

As series editor, I have been working closely with authors to produce volumes that capture the spirit of a user's guide that gives the reader enough information and advice to successfully conduct a project using each method. I hope I have not been an annoying gnat in their ear. Fortunately, I have chosen authors who are not only talented but also patient with a meddling series editor.

The most important job of a series editor is to pick the right people to produce the right topics. Together with Lois Smith, Steve Stafford, and others in the HFES Communications Department, HFES Executive Director Lynn Stroether, the advisory board, and the chair of the Publications Division, Melody Carswell, we were able to gather input through surveys and contacts to choose the topics. I hope to continue periodic surveys so that timely topics can be added to the series.

For this first volume on usability assessment, we invited Phil Kortum from Rice University to be the adviser over our shoulders. I think you will agree that Phil is a nice voice to add to your head. I also want to thank Phil for working with me as we shaped his volume and, as a consequence, the rest of the series.

Francis T. (Frank) Durso
August 2016

PREFACE

One hundred years ago, if you wanted to compose a letter, all you needed was a pencil and a piece of paper. Mastery of those simple tools was all that was required to begin the task of writing. Wanted to read the news? Purchase a newspaper and open it. Even more difficult tasks, such as doing your taxes or solving mathematical equations were done with paper and pencil.

In today's technologically complex world, things are not so simple. Computers and their associated programs are used extensively in our daily lives. We write our papers on word-processing programs, read the news on the Internet, and solve mathematical equations using spreadsheet programs or symbolic math programs. Indeed, in today's world it is not usually sufficient to know only what you want to say or how to solve an equation; you also have to be able to master the *tools* that enable you to perform those tasks. This is why the usability of those tools has become more important than ever. Products and services that are unusable decrease the user's efficiency, lower their satisfaction, and, in many cases, even lead to injury or death. More to the point, products that are unusable typically *don't get used.*

Marketing folks might disagree, but it's insufficient to simply state "this tool is easy to use." You can say it, but if it's not true, your users and buyers will quickly discover that fact. If you are developing or purchasing a new product for your company, you need to be able to quantify how usable a specific product is so you can benchmark it against other products or improve its usability if it's not up to par.

Although there are many excellent texts on how to assess the usability of products, services, and systems, I think you'll find this book a valuable addition to your library, particularly if this is your first foray into the world of usability assessment. As the first in the Human Factors and Ergonomics Society's *Users' Guides to Human Factors and Ergonomics Methods* series, this book has been written to help the practitioner by presenting the material clearly and succinctly, with copious examples.

If you've never done a usability assessment before, fear not. This book is specifically designed to guide you step by step through the process of performing and reporting the results of a rigorous usability assessment. It is my hope that students and practitioners alike will find this book to be a valuable resource as they strive to make their products and services as usable as possible.

Acknowledgments

I thank my editor, Frank Durso, for inviting me to write this book and for trusting me to be the first author in this new HFES book series. Thanks must also be extended Carol Stuart-Buttle, Kevin Gildea, Jeff Sauro, and Bruce Walker, who so generously provided kind but thorough comments that have made this book the best it could be. In addition, I thank Claudia Acemyan, a former student who helped review some of the first drafts and provided invaluable feedback from a student's perspective. The efforts of the folks at HFES, notably Lois Smith and Lynn Strother, who helped shepherd the book to fruition, are also greatly appreciated. Finally, thanks to the thousands of people who have participated in my usability assessments over the years, and who inspire me to continue to work toward designing the most usable systems possible.

CHAPTER 1: WHAT IS USABILITY ASSESSMENT?

1A) A Brief History

Usability is about making things easier to use. Whenever we use a product, service, or system, we want to be able to perform our tasks quickly, with a minimum of errors. For most products, we want to spend little, if any, time learning to use the product. Such is our basic concept of usability.

Using this simple definition, it could be argued that the fundamentals of usability have been around since the invention of tools. An early human user might have decided that modifying a hunting tool would enable him to make fewer errors when using it, to hunt with it more efficiently, and to be happier with it overall. This early usability assessment of the hunting tool likely would have led to greater success in this critical endeavor.

Although we might trace the origins of usability back to the dawn of humans, the advent of usability assessment is typically thought to have its beginnings when modern technologies began to emerge. As modern technology became prevalent, weaknesses in human perception and cognition became more evident as we tried to learn and operate these increasingly complex systems.

The concept of usability, also referred to as *human factors*, began to take hold during World War II, as the idea of finding a specific person who could do a particular job was replaced by the idea of designing jobs so just about anyone could do them. This new concept was necessary because of the need to have any person do any job in furtherance of the war effort. One of the most famous examples can be found in Alphonse Chapanis's study of aircraft landing accidents.

In early models of the B-17 bomber, pilots making an approach for landing would raise the landing gear right before touching down, crashing the aircraft. The Army Air Corps classified these incidents as "pilot error" and brought Chapanis in to see if better training methods could be developed. Chapanis noted that these accidents were not occurring on other bombers, whose pilots were similarly trained, and began an assessment of why this might be happening.

After observing pilots using the aircraft, he discovered that the controls for the flaps and the landing gear were nearly identical in shape and located in close proximity (see Figure 1A). In high-stress conditions such as landing, pilots confused these two controls and ended up raising the landing gear rather than extending the flaps (Chapanis, 1999). His suggested design, shown in Figure 1B, persists today in even the most modern aircraft. In that design, the landing gear control looks and feels like a wheel, allowing pilots to easily find and identify the control, even in high-stress situations.

After the war, usability began to move into more industrial settings. In 1947, John Karlin founded the User Preference Department at Bell Laboratories, working on projects to understand how users interacted with telephone systems. His group examined the transition from exchange dialing (e.g., Pennsylvania 65000) to all-numeric dialing, and they helped determine which configurations of touch-tone telephone keypads were easiest to use by people who were familiar with rotary dialing (Hanson, 1983).

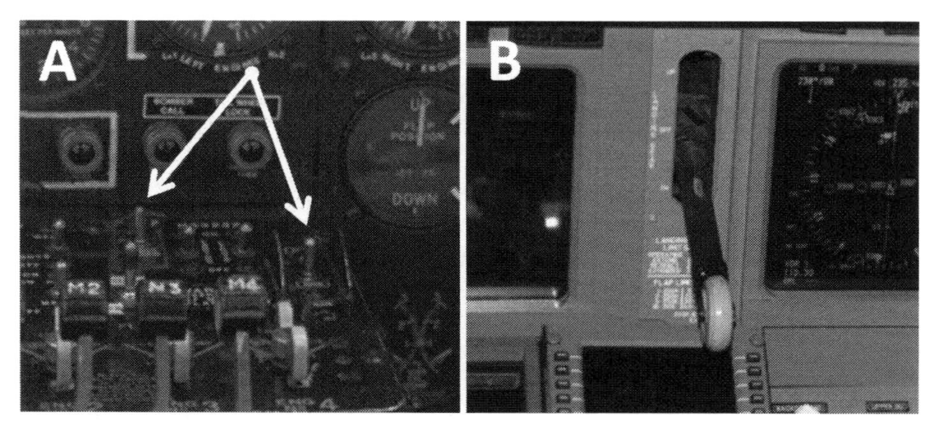

Figure 1: **(A)** The original configuration of the B-17 landing gear and flaps control. Note that this is a later version of the cockpit that had already been modified with a guard to help pilots avoid making the error *(photo modified from Darren Kirby, used by permission under CCASA 2.0).* **(B)** The landing gear control on a modern Boeing 727 aircraft *(photo modified from Politikaner, used by permission under ShareAlike3).*

In the late 1970s and early 1980s, usability became increasingly important in the health and safety domains. A number of high-profile accidents occurred that were caused by usability issues associated with complex technological systems and computers, which were becoming more and more ubiquitous. The near meltdown at the Three Mile Island nuclear power facility in Pennsylvania (Malone et al., 1980), the death of patients from massive radiation overexposure with the Therac-25 medical imaging device (Leveson & Turner, 1993), and a number of prominent aircraft accidents resulting from the poorly understood usability of their systems (e.g., Wiener, 1985) led to an increased interest in usability assessment.

This era also saw the advent of the first home computers, the rise of the graphical user interface (GUI), and the beginnings of the widespread adoption of the precursor to the Internet (ARPAnet). These new computer systems taxed users' cognitive and perceptual abilities in ways that had not been common before, and the deleterious results were not always anticipated. These usability limitations helped contribute to such incidents as the large Dow Jones decline attributed to a single broker's computer input error (Hurtado, 1992) and the crash of an Airbus 320 during an air show after the pilot turned off the flight safety systems because he didn't appreciate their importance (Casey, 1998).

Today, we are also concerned with the usability of mobile devices, which have become nearly indispensable tools in everyday living. The usability of these mobile devices and in-car electronics has gained significant attention, as nearly 27% of accidents are now being attributed to the use of these devices while driving (National Safety Council, 2013).

New technology always promises the user that they will be able to perform their tasks quickly and easily with a minimum of effort. History has shown us, however, that without rigorous usability assessment, these claims can ring hollow. Although much of this lack of usability simply leads to frustration and decreased productivity for the user (who among us has not struggled with our word-processing program at some point?), other instances of poor usability have led to the loss of life and limb, as illustrated in some of the foregoing examples.

1B) What Makes Something Usable?

Something is usable if users can accomplish their goal in a way that makes sense to them, and they can do so with minimal effort. The usability of simple items, such as the bottle opener shown in Figure 2A, is fairly easy to determine. If I can pick up the opener and quickly open my ice-cold soda, the device is usable. The usability of more complex items, such as the digital watch shown in Figure 2B, can be more difficult to ascertain, but it's still relatively straightforward because the watch has a few fundamental tasks that must be accomplished, such as setting the time and the alarm.

Determining whether a significantly more complex system is usable, like the aircraft cockpit shown in Figure 2C, can be much more difficult. Not only can a large number of tasks be accomplished, but there are also numerous subsystems that may be involved in completing any given task. When we think about extremely large, complex systems that are made up of many different subsystems with different functions, such as the combat ship shown in Figure 2D, determining if such a system is usable overall can be quite involved. Which systems are involved in such an assessment (e.g., weapons, navigation, communication), and how do we weigh their relative importance in the overall assessment of usability?

It has been suggested that products and systems with complex, multiple, simultaneous goals can't be reliably assessed from a usability standpoint (Bevan & Raistrick, 2011; ISO 20282, 2006). Although the complexity of the system can add time and expense to the usability assessment, the basic methods used are generally the same. More important, the fundamental attributes that make up a usable system are the same regardless of system complexity.

In general, usable products, services, and systems have six important attributes:

1. They are useful and perform a specific desired task that the user wishes to accomplish.
2. They can be used efficiently to accomplish the desired task in a timely manner.
3. They are effective and enable a user to accomplish a given task without error.
4. They are easily learnable and can be used with little or no instruction. Obviously, very complex systems such as an aircraft cockpit can't be used without instruction. However, when properly designed, even such systems have generally favorable learning properties.
5. They are generally very satisfying to use.
6. They are accessible by persons with disabilities.

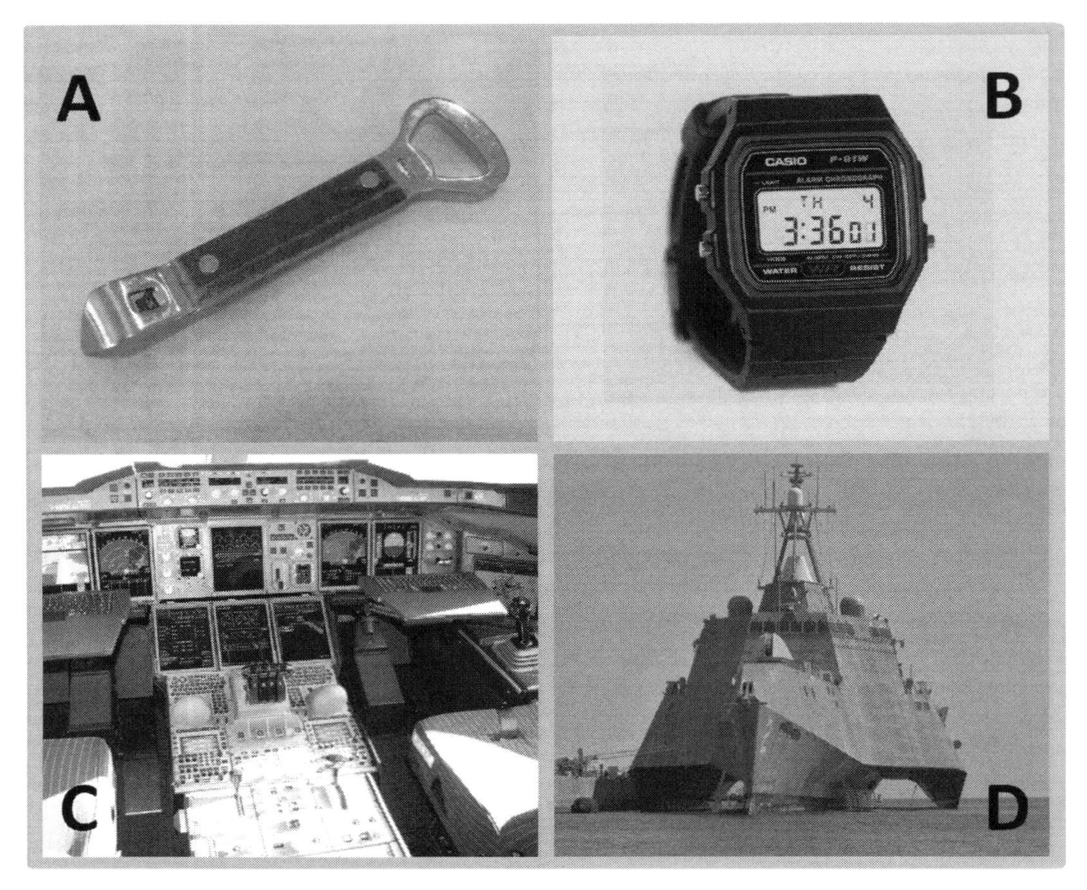

Figure 2: Examples of systems of varying complexity that undergo usability assessment to determine their overall ease of use. **(A)** A simple bottle opener *(image copyright Philip Kortum)*. **(B)** A typical digital watch *(photo by NotFromUtrecht, licensed for use under CC by 3.0)*. **(C)** A modern aircraft cockpit *(photo by Naddsy, licensed for use under CC by 2.0)*. **(D)** The U.S. Navy's newest Littoral Combat Ship, an extremely complex system of systems *(U.S. Navy public domain image)*.

1C) Myths of Usability Assessment

Over the course of many years, I've had the opportunity to hear all manner of excuses as to why usability assessment isn't required, doesn't add value to the process, or isn't necessary for a particular product or service. It's not that people who repeat these myths and are involved in the product development process don't want the product to be successful (or usable). Obviously they do. However, because of cost and time constraints, these team members are often under significant pressure to push the product along in its development path, and they can often see usability assessment as an impediment to on-time/on-cost completion of the task. Usability professionals know better and should tirelessly work to ensure that usability assessment is performed to help ensure the greatest chance for product success in the field.

Here are seven of the most common myths about usability assessment.

Myth 1: I'm a human, I can use it; usability assessment complete.

This particular myth is most commonly spoken by programmers. Having worked diligently and expertly on the product, they have developed robust mental models of how the system works and have become experts in its use. They imagine that all users will come to the product with this expertise or that the product is so intuitive that no expertise is required. This is not to impugn the skill or dedication of these programmers. Their code is tight, their algorithms are eloquent, and their implementation can be flawless. However, it is likely that they have not accounted for the broad range of users who will ultimately use the product.

There's an old saying in usability assessment circles: "You are not the user." What this means is that although you may be a user, you're not the only user, and so your experience may or may not be typical of the rest of the population of users. Even if your experience and expertise are similar to those of other users who will ultimately use the product, one user (you) is insufficient to make an accurate assessment of the usability of any given product or service. Even if you *were* the only user of the product (a highly unusual situation), you likely wouldn't have a full understanding of how the product would function under user stress, user fatigue, unusual or unexpected situations, and long-term use without going through a more formal usability assessment.

Myth 2: Usability assessment is just common sense.

This misconception is likely driven by the simplicity of the examples that are often used to describe human factors and usability principles. For example, the principle of consistency may be described with an example of a bathroom with two sinks, where the faucet handles on one sink turn clockwise and the faucet handles on the other sink turn counterclockwise. An average observer might look at this situation and exclaim, "What were they thinking? Common sense would dictate that the knobs should turn the same way."

In another example, the principle of affordance might be described by a door with a pull handle that actually pushes open. Again, the average observer would examine the situation and say, "It's just common sense not to have a pull handle where you want to push." Indeed, in both of these situations one could argue that the design failures might be described as a lack of common sense.

However, in many other usability issues, common sense isn't common at all. Take, for example, the length of a personal identification number that is required to access some secure device, like a bank password on a Web site. A security expert will say that the longer the code is, the better, and might suggest a number that is 50 to 100 digits long. However, we all know from personal experience that memorizing a 100-digit number would be difficult. What *is* the maximum length that would be easily usable by a human? Common sense does not help us here; instead, we must rely on psychological data and usability assessment to help us understand how to design interfaces that support human memory limitations.

Let's think about another example, where we are trying to design a usable interface that uses color to maximize the likelihood that a user will detect a specific feature, such as an emergency cutoff switch at a gas station. What does common sense tell us about which color has the greatest detectability? Again, common sense does not help us here. We must rely on physiological data that tell us what the sensitivity of the human visual system is to different wavelengths of light. Only then can we select a color that will be easily detected.

Interestingly, in this case common sense has proven to be wrong. An example of this commonsense failing can be found in the colors that we paint fire trucks. One desirable trait of a fire truck is that it is visible so people can see it, get out of its way, and not interfere with its task of getting to the fire. Most fire trucks are painted red, and so, using common sense, we might say red is an excellent color for high detection of a fire truck. Further, fire is red, and so there appears to be a good mapping between the color of the truck and its purpose. Unfortunately, these intuitions are wrong. It turns out that red is one of the most difficult colors to detect, particularly at night. Instead, lime green colors should be used because they maximize detection capability by matching the peak sensitivity of the human eye to these wavelengths, a finding backed by empirical evidence (Solomon & King, 1997).

As can be seen, there's nothing "common" about common sense. Instead, you must rely on empirical evidence to make good usability judgments.

Myth 3: My system will use highly trained specialists, so assessment isn't necessary.

This argument is most often heard in large organizations, like the military or Fortune 500 companies, where there is a significant and mature training apparatus for the users. The suggestion is that any deficiencies in usability can be made up for with specific and rigorous training. This may be true to some degree, as education and experience can help mitigate some aspects of poor usability. However, experience has shown that even the most highly trained individuals can continue to make egregious errors in the field if systems and products have usability deficiencies.

For example, I think most people would agree that physicians are among the most highly trained individuals in the world. That said, the Institute of Medicine estimates that approximately 98,000 people in the United States die each year from *preventable* medical error (Kohn, Corrigan, & Donaldson, 2000). More recent estimates (James, 2013) put that number closer to 400,000. Other industries, such as nuclear power plant operations and air transportation, also employ some of the most specialized and highly trained individuals. Again, this training has not prevented serious usability-related failures (e.g., Three Mile Island).

The hallmark of well-designed, usable systems is that they require minimal training. In fact, some the best systems may require no training at all – they are simply "walk up and use."

Myth 4: My system is fully automatic, so no assessment is necessary.

With the advent of modern computers, we have ceded more and more control of systems to the machine. As a result, many designers believe that usability assessment is unnecessary. If the computer is taking care of everything, why would usability be important? The designer's goal in this case has been to remove the human from the loop under the assumption that if a human isn't in the loop, usability assessment becomes unnecessary.

These designers forget that even in fully automatic systems, users must interact with the system at some point. People have to set the systems up, maintain them, and, most important, interact with them when the system has failed to operate in the expected manner. In this instance, exceptional usability of the systems becomes paramount.

Consider the example of China Airlines Flight 140. It was en route from China, on final approach to its destination at Naoya Airport in Japan. During the flight, one of the crew had inadvertently activated the takeoff/go-around button in the cockpit. This is a piece of automation that assists the pilots when they are trying to take off or abort a landing, to prevent them from crashing into the ground. In this case, however, the pilots were actually trying to land the aircraft.

As the pilots pushed down on the control yoke to land the aircraft, it actively pushed back, adjusting the engine and pitch to prevent the aircraft from landing (or crashing, in its judgment). In the end, this battle between aircraft and crew caused the plane to stall, ending in its crash and the loss of 264 lives (U.S. Department of Transportation, n.d.). In this case, the automatic system went against the goals of the user and, more important, didn't communicate to the user about *its* goals. The pilot couldn't figure out why the plane wasn't behaving as expected, and the automatic system that was flying the plane wasn't usable enough to communicate to the pilot what was happening, with tragic results.

Scenarios like these will likely become significantly more common as self-driving cars become the norm and these kinds of complex automated systems are used by nearly everyone, not just highly trained pilots.

Myth 5: My system isn't mission-critical, so the time and expense isn't worth it.

When highly complex systems are being developed, we expect (hope?) that usability assessment will be part and parcel of those efforts. It is easy to understand that if we are sending humans into space, it's probably very important that the space capsule be thoroughly assessed for usability in this mission-critical setting. We assume that extensive usability assessments have taken place for the latest Air Force fighter jet, given that small decrements in user performance could have significant consequences in a battle situation. However, as we move toward more consumer-grade products, usability assessment is often seen as a "nice to have" rather than a "must-have."

There are certainly exceptions. Some consumer-product-oriented companies have positioned themselves as leaders in highly usable products, and much of their marketability is based on the fact that their products are easier to use than those of their competitors. Think of the iPhone. In this case, usability assessment enjoys nearly the same status as it does in companies that build products for the space or defense industries. Many times, though, designers see their products as insufficiently mission-critical enough to warrant the additional expense required by rigorous usability assessment. They believe that small decrements in user performance, though unfortunate, do not rise to the level of criticality that would be required to justify the time and expense that is needed to perform the level of usability assessment that is generally accepted in mission-critical systems.

The fact is, if someone is using your system, it probably is mission-critical to *them*. More important, your system may function as part of a larger set of systems, and usability deficiencies in your product may have larger consequences than one might imagine at first blush. For example, let's say you are developing the latest iPhone application. Budgets are tight, so you decide that even though the application is a little bit less usable than it could be, the extra few seconds that it costs your users to operate your new app aren't really worth the added expense on the front end of the development process to iron out those weaknesses.

What happens when your users try to use your new app as part of another mission-critical activity, like driving? If you fail to anticipate situations in which your application would be used as part of a larger system in which the mission-criticality of the user performance of *your* component plays a part, you haven't adequately considered all the important dimensions of usability. Although consideration of every possible scenario in complex systems isn't always possible, the practitioner should at least try to anticipate *likely* scenarios.

Generally, if your product is important enough to build and sell, it's important enough to have rigorous usability assessments conducted as part of the development process. More important, good usability assessment doesn't have to be expensive and time consuming, so there's no excuse not to do it.

Myth 6: My system uses off-the-shelf components, so no usability assessment is needed.

Often, the interface of a system is simply a collection of subcomponents, each with its own interface, that have been integrated into a whole unit. Designers often make the mistake of thinking that the usability assessments performed on each of the subcomponents guarantees that the resulting integrated system will have favorable usability characteristics. However, even if each of the components has perfect usability characteristics on its own, once they are collected into a new system, the usability characteristics of that new system are now unknown.

Think of a hospital operating suite. Inside that suite are large numbers of systems, such as infusion pumps, ventilation or anesthesia delivery devices, and monitoring equipment that must function together in order for a surgeon to successfully operate on a patient. Each of these myriad systems has its own alarm system and, according to FDA regulations, has undergone rigorous usability testing to ensure that operators of each specific piece of equipment can hear, understand, and act on that alarm.

Imagine, however, a situation in which things go badly during surgery and all of the alarms are now sounding simultaneously. The surgical team becomes overwhelmed with warnings, and the cacophony of sounds makes it difficult to sort out which needs are most important and how best to deal with the impending crisis. Although each individual component might operate flawlessly from a usability standpoint in this regard, collecting them into a system has created a new set of usability problems that weren't anticipated at the component level. Therefore, it is critically important that the assembled system of components is tested in the state in which they are expected to be used. This enables designers to understand the interactions of the components and how usability might be adversely affected.

Myth 7: No real users would be stupid enough to use my product in that way.

Sadly, some product teams refuse to believe that their products could be difficult to use. They would rather blame the user, making the assumption that the tested population simply couldn't be the population that's going to use their actual product. Even after watching many users fail in their attempts to use the product, they'll simply say, "My users won't be that stupid." If you've done your job correctly and set up a good, rigorous test with a representative user population, poor results found during the usability assessment are likely to accurately predict poor usability results in the field.

Generally, users are not stupid. They have a goal, and they are attempting to use your product or service in a way that accomplishes that goal. If they perform actions that cause the product to fail, it's likely that they thought their actions would result in success, not failure.

Users often bring a different mental model of the operation of a product to the task, and their attempts to use your product are consistent with *their* mental model, not that of the designers. As can be seen in Figure 3, the designers of this security card entry device for the entrance to Rice University's main library clearly had a different mental model of how the system would work than the average user, and the library staff finally had to augment the interface to assist the users.

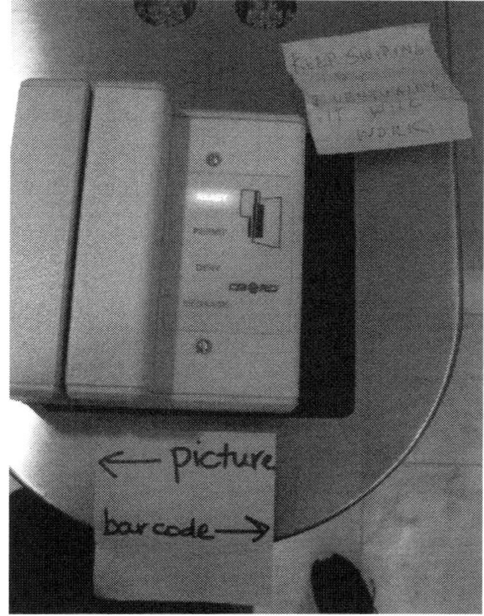

Figure 3: Users of this interface consistently misused the device, not because they were stupid but because their model of how it worked failed to match the model of the designer. A simple usability assessment would have likely found this deficiency and enabled the designers to make changes before fielding the product *(image copyright by Philip Kortum).*

Sometimes designers choose aesthetics over usability, believing that users will quickly adapt and appreciate the design. A good example of this is shown in Figure 4A, the paper towel dispenser in the men's restroom at Austin Bergstrom International Airport when it first opened. To their credit, the designers quickly discovered that users were having a difficult time finding the paper towels and installed the small sign shown in Figure 4B. Unfortunately, this modification was insufficient to aid the users, and eventually the designers (again, to their credit) installed a conspicuous sign, as shown in 4C, to help the users find the paper towels.

Clearly, the aesthetic design of the original paper towel holder has been completely lost with the continued modifications that were required to help users achieve the simple goal of drying their hands.

Although users are not stupid, it is true that they are exceptionally creative in how they will use products to try to achieve their goals. If the system provides users with opportunities to make errors, you can be certain that errors will be made.

Remember, the goal of usability assessment is to find and fix usability deficiencies in the laboratory, before the products are launched, to minimize human error that results in monetary loss, customer churn, failure to achieve the desired outcome, or injury or death of the user.

Figure 4: Poor usability leads to multiple iterations to support user goals. **(A)** The original paper towel holder design (that's the paper towel in the center of the picture, just below the seam in the mirrors). **(B)** The modified design, with a small sign to direct users to the towels and **(C)** the final modification, with a large sign to help users accomplish their simple goal *(image copyright by Philip Kortum).*

1D) Suggested Reading

- *A History of Usability* by Jeff Sauro. Online at http://uxmas.com/2013/history-of-usability.
- *The Design of Everyday Things* by Don Norman. Basic Books, 2013.

CHAPTER 2: WHY ASSESS USABILITY?

2A) Product Improvement

Generally, most users simply don't think explicitly about usability. They want to perform a task using your product or service, and they expect that they should be able to do so without any degree of difficulty. This leads to a universal usability truism: *Good usability is invisible.* Users are often captivated by the clever engineering that it might take to make a product work, or the artful styling that differentiates the product from its competitors. However, when it comes to the usability of the product or service, unless there is a problem, users assume that this is how the product – or any product like it – should work. Only when failures occur do users become concerned about the usability of the device.

Consider a review of the latest sports car. The writer will rave about the engineering that went into the responsiveness of the engine and the superb handling. She'll extoll the artful styling and describe the emotional experience of viewing the car. When it comes to the interface and its usability, however, she'll simply say, "The controls were where *they were supposed to be*" indicating that although the mechanical engineering and industrial design aspects of the automobile took considerable skill, the superior usability of the design happened because there really wasn't any other way to do it. In fact, it is likely that as much effort went into the human factors and usability aspect of the design as was devoted to the mechanical aspects of the automobile.

Even though good usability may be invisible, poor usability can have significant consequences. The goal of usability assessment is to improve the product and minimize or mitigate these consequences. This product improvement generally takes place in four areas:

1. Reduction in user error
2. Increased user efficiency
3. Increased safety, for both the system and the user
4. Increased user satisfaction

2Ai) Reduction in User Error

Error can be described as actions taken by users that reduce the effectiveness, safety, or performance of a system (Sanders & McCormick, 1993). Note that this definition does not require an adverse consequence – simply the opportunity for an adverse consequence to occur is sufficient for an event to be classified as an error. Highly usable systems help reduce the cognitive load of users, which, in turn, results in the commission of fewer errors by those users. A reduction in user errors is the "Holy Grail" in mission-critical systems. By decreasing the number of errors that users make, you can increase both the effectiveness of the system (the ability of the system to do what it was designed to do) and the efficiency with which people can operate the system.

Aren't usability assessment and quality assurance testing the same thing?

No! Usability assessment uncovers errors that users might make while using the product, whereas quality assurance testing looks for software coding bugs that might prevent users from performing the tasks that they wish to complete. Quality assurance testing systematically examines every feature and function of the product to ensure that it actually exists and that it operates as specified in the requirements document. Usability assessment focuses on the most common tasks that users perform and tries to determine if the interface elements of the product contribute to user errors. Although usability issues are sometimes found in quality assurance testing, that is not its primary goal. Both usability assessment and quality assurance testing should be conducted to ensure that the product is of the highest possible quality.

Another aspect in the reduction of user error is the concept of error recovery. We know that users are going to make errors (despite our best usability efforts), and so often the question is not how to completely prevent the error from occurring but, rather, how to best provide mechanisms to assist a user in identifying the commission of an error and then gracefully recovering from it.

Figure 5 provides an excellent example of what *not* to do. Imagine that you've been working on an important paper and accidentally hit a keystroke sequence that selected the entire document and deleted all of its text. In your haste, you hit the "save" button and you get the message shown. The action allowed by the message is the *one* action that you don't want to take, but the interface does not provide you with any alternatives. In this case, the user has been able to successfully identify that he has committed an error, but he's unable to recover from it. A more usable interface would simply add a button that enables the user to decline the overwrite action.

By allowing graceful error recovery, we can mitigate any adverse consequences that might occur despite our best usability efforts.

Figure 5: A good example of a dialog box that has poor usability because it does not allow for the graceful recovery from the underlying error *(image by Philip Kortum).*

Sometimes poor usability can have significant consequences, and the need for reduced error is apparent. Consider the case of a juvenile defendant who was able to smuggle a loaded gun through the security screening checkpoint at the Travis County Courthouse in Austin, Texas, even though he entered the facility through a metal detector. It turns out that the metal detector had been unplugged *for three years,* and security personnel were unaware that the device was not functioning (Rogers, 2009a, 2009b). Obviously, in these kinds of security situations, the ability to determine that the system is operational is a fundamental usability aspect of the device. Given the duration of the failure, it seems miraculous that this story did not have a tragic ending.

In case you might be tempted to invoke the usability myth that "my users would never be that stupid," three years after this incident, an entire terminal at JFK airport in New York and the passengers from two planes that had already pushed away from their gates had to be evacuated and rescreened because one of the metal detectors at the security screening checkpoint in Terminal 7 had been unplugged for the entire morning and the TSA agents had not noticed (Messing, 2012).

2Aii) Increased Efficiency

Systems that are highly usable are also more efficient to use. The user can easily move through each of the steps in a task and not have to devote additional time to figuring out how to perform the next step. Further, because users are making fewer errors, they can spend less time back-tracking through the interface and recovering from unnecessary errors.

Adverse consequences associated with a loss of efficiency arising from poor usability are wonderfully illustrated by the 1977 blackout in New York City. A lone power plant operator was tasked with managing the city's electrical grid, and much of the city's electrical supply had been compromised by lightning strikes on transmission and generating facilities in the upper northeast. The Con Edison operator needed to bring New York's own in-city generation facilities online and shed some electrical load to in order prevent the entire grid from going offline.

Unfortunately, the displays that would have informed the operator of the overall state of the system were poorly designed and spread across several rooms, so the operator never could gain a clear picture of how dire the situation was becoming. His inability to act in a timely fashion because of the system's poor usability led to the complete collapse of the New York City power grid, plunging the city into darkness for several days and leading to widespread looting and arson (Casey, 1998).

In some critical situations, efficiency can mean the difference between life and death. If I have a hard time figuring out how to use a defibrillator, that lost efficiency can lead to a failure in the task of saving a life. In most normal situations, especially for consumer products, the efficiency gains made with highly usable products and services simply enable a user to complete the task in a minimal amount of time, thus increasing his or her productivity and leaving more time for the completion of other tasks.

2Aiii) Increased Safety

One significant outcome of these decreased errors is that highly usable systems can be safer, resulting in the reduced likelihood of injury to both the user and the system. If the system is easy to use, users can focus on the appropriate tasks at the appropriate times. Usable systems result in users making fewer inadvertent errors that could cause unsafe conditions.

Poor usability, on the other hand, can lead to significant undesirable consequences. In 2002, the Children's Hospital of Pittsburgh implemented a new computerized physician order entry system that was supposed to significantly improve how patient information was collected and shared. Unfortunately, the system was fraught with usability issues: Doctors couldn't enter medication orders in a timely fashion, nurses couldn't access critical patient information when needed, and medical teams spent more time trying to figure out the system than they did examining their patients. The result of this poor usability was that child mortality at the hospital *more than doubled* after the system was implemented (Han et al., 2005).

Two thousand miles away, patients at Cedars-Sinai Hospital in Los Angeles were turning up with very odd hair-loss patterns after routine brain scans. For 18 months, doctors and technicians had failed to note that the system had been set to deliver up to 10 times more than the safe dose of radiation. As a result of this lack of usability, more than 250 patients suffered from significant radiation overdoses before doctors connected the dots and figured out what was happening (Bogdanich, 2010).

In both of these cases, poor usability led to situations that were inherently unsafe. Hundreds of patients in Los Angeles will have to live with a significantly increased chance of getting cancer for the rest of their lives, while in Pittsburgh there was a significant loss of life. Why didn't the computer system at Children's Hospital save lives rather than contribute to children's deaths, and why didn't the CT machine in Los Angeles simply refuse to perform an action that it should have known would harm someone? Rigorous usability assessments of these products might have identified these deficiencies and prevented the resulting dangerous situations from occurring.

Often, the results of usability testing suggest that a simple design fix could rectify the usability deficiency. Take, for example, the case of the inadvertent ignition of a rocket that was still in the assembly hanger. A technician was assigned to perform a continuity test of the ignition system. As was standard practice, the technician replaced the battery on the test device to ensure a reliable power source. Unfortunately, the battery tray, whose function was to significantly reduce the voltage delivered to the test set, wasn't replaced properly, and when the continuity test was performed there was sufficient voltage to ignite the rocket motor. This resulted in the loss of one life and a multimillion-dollar rocket system (Casey, 1998), which could have been prevented if the battery tray had not been removable.

2Aiv) Increased User Satisfaction

Consumers want to be satisfied with the products they use. There's nothing worse than using a product that you hate to use. From a marketing standpoint, highly usable products result in several important benefits.

First, users who are satisfied with the current version of the product are more likely to purchase future versions of the product. They're also more likely to integrate that product into their daily activities. A single satisfied user can also serve as a tremendous marketing asset by recommending the product or service to friends and professional colleagues, which is especially important in today's hyper-connected social media world.

The iPhone is an excellent example of the importance of usability in creating high user satisfaction (Lee, Moon, Kim, & Mun, 2015). This device is perceived as having ease of use that is superior to its competitors, and that ease of use has resulted in extremely high user satisfaction. It has also created exceptionally loyal users who repeatedly purchase not only the iPhone but other related Apple products and frequently make recommendations of the entire line of Apple products.

Does satisfaction play a role in systems that users *have* to use? For example, a pilot rarely gets to choose the plane she'll fly, and an engineer working for a large corporation will be given a computer with the design software that the company typically uses. Even in these cases, increased user satisfaction is an important benefit of highly usable products. Users will likely be more effective and efficient (because they're committing fewer errors) when using products that result in high satisfaction. More important, users are more likely to actively adopt and use systems that have been demonstrated to have high usability (Holden & Karsh, 2010).

Clearly, the concepts of error reduction, safety, efficiency, and user satisfaction are related (Sauro & Lewis, 2009). Assessing the usability of your products and services and then taking the appropriate action to rectify the deficiencies identified during the assessment will prove beneficial in all four areas.

2B) Financial Benefits

Although increases in safety, customer satisfaction, and reduced user error are important in their own right, all three can be monetized, and this is another important reason that usability assessment is critical. Let's face it; even though every company wants its products to reduce the number of errors a user makes, increase the safety of those users, be highly satisfying to use, and be seen as of the highest quality, it all comes down to cost. Businesses are supposed to make money, and to the extent that highly usable products contribute to this goal, usability assessment has a place.

Studies have shown that for every dollar a company invests in usability efforts, it will realize between $10 and $100 in return (Gilb, 1988). The financial benefits of fielding highly usable products can come from increased sales, reduced customer churn, reduced support costs, and reductions in the financial consequences related to product failures that result in the loss of life and limb or equipment (Marcus, 2005). Because consumers are goal oriented, they are likely to purchase and continue to use products and services that help them achieve those goals. When customers are dissatisfied with their ability to perform a task with a product, they tend to seek an alternative. Customer churn can be a major expense for large companies, and so reductions in customer loss caused by increased usability can be highly beneficial (Gupta & Zeithaml, 2006).

Determining the fraction of your sales and customer churn that are directly related to usability can be difficult, but mapping increased usability to reduce support costs is much more direct. If a product is highly usable, fewer of your customers will require support services. If you have a large user base, the savings can be substantial. Let's say you have 10 million customers (that's the number of people who currently subscribe to a popular television cable service in the United States), and every month 10% of those customers have to contact your support services because of usability issues. If we assume that each call costs $25 (Rumburg, 1998), your company is incurring approximately $25 million a month in direct support costs. If we are able to improve the usability of the product so that just 10% fewer people have to call the support center, the annual savings is $30 million. Clearly, as your customer base grows, small decreases in the numbers of people who have to use your support service can lead to large monetary savings.

Sometimes those cost savings can come from changing the paradigm of customer support from a high-cost option to a lower-cost option. In the case of DSL high-speed Internet self-installation, which is described in greater detail in section 3, the goal of the company was to stop dispatching technicians to perform the DSL installation and instead have users perform their own installation. Sending a technician to a customer's home can cost between $200 and $2,000 (McPherson, 2012). If we make a conservative estimate and say that a DSL truck roll costs $500 and the company is installing 1 million DSL lines a year, that could result in an annual savings of $500 million. Even if we assume that *every single one* of those customers has to make a $25 support call, we've still saved over $475 million simply by shifting the paradigm from a truck roll to customer self-install. As you'll read later, that paradigm shift was substantially supported by making the self-installation kit usable enough for customers to perform their own installation.

In several earlier examples, I've shown that poor usability has the potential to lead to the loss of life and limb. However, that lack of usability can also lead to the loss of the product as well. There's a reason the military spends a lot of time and money on usability assessment: They know that small decrements in usability performance can lead to large losses. If the usability of my opponent's aircraft is better than the usability of my aircraft, he might be able to shoot a missile at me before I can shoot a missile at him. Even if I can eject and live to fight another day, I've just lost a $65 million aircraft (the estimated cost of the new F-18 Super Hornet). In other cases, failure can lead directly to large-scale loss of life. Companies may then be motivated by the significant financial burdens resulting from legal actions surrounding the failure.

The crash of American Airlines Flight 965 in Cali, Columbia, serves as an excellent example. The crewmembers were trying to select a specific navigational beacon on their Boeing 757's flight navigation computer, but the system auto-selected another similarly named beacon instead. When the plane's autopilot began to fly toward that beacon, the crew focused all their attention on why the plane was not reacting as they expected. In the ensuing confusion, they lost all situational awareness and crashed the plane into a mountain, killing 260 of the 264 passengers aboard. Damages in the amount of $300 million were shared by American Airlines and the manufacturers of the navigation equipment and its software (Morrow, 2000).

Although avoidance of potential liability issues may not seem like the most altruistic reason to perform usability assessment, these kinds of potential liability costs can certainly help companies recognize that usability efforts can have significant value.

Is it really worth it?

Absolutely! If you are testing mission-critical systems in which errors or mistakes could have potentially devastating consequences for life or limb, usability testing is essential. Numerous stories are told describing simple usability failures that led to the loss of many lives (see Casey, 1998 and 2006, for many excellent examples), or resulted in the potential for catastrophe (e.g., Three Mile Island). In most of these cases, one is left to wonder how the designers of the systems did not foresee that these kinds of simple errors could lead to the resulting devastating consequences.

The answer is that sufficiently rigorous usability testing was not conducted. Even in systems in which life or limb is not in danger, mission-criticality remains. If I run an e-commerce site, the ability of users to browse, select, and purchase my product *is* mission-critical. Failure to understand the kinds of errors that prevent users from performing these simple tasks can cost your company millions of dollars.

Although usability is often seen as "nice to have," it is actually critical. If customers can't use your product, you can be certain that there will be adverse consequences. Those consequences could be as simple as a loss of reputation and revenue as consumers choose your competitors' products over yours. Or they could be as severe as significant injury or death to not only the direct users of the product but innocent bystanders as well, leading to moral and legal liability. In either case, the cost of usability testing is well worth the investment.

2C) Suggested Reading

- *Cost Justifying Usability: An Update for the Internet Age* by Randolph Bias and Deborah Mayhew. Elsevier, 2005.
- *Usability Is Good Business* by George Donahue, Susan Weinschenk, and Julie Nowickie. Online at http://half-tide.net/UsabilityCost-BenefitPaper.pdf.
- Usability and the bottom line by George M. Donahue. *IEEE Software, 18*(1), 31–37, January/February 2001.

CHAPTER 3: PREPARING TO PERFORM THE USABILITY EVALUATION

3A) Introduction to the DSL Self-Installation Example

The next few chapters will take you through the nuts and bolts of performing a usability assessment. I describe the steps necessary to perform a rigorous usability assessment of any given product, service, or system, from defining your users all the way through reporting the results.

Figure 6 shows the three phases of usability assessment and the chapters that encompass them. Chapter 3 will help you prepare to run the assessment. Chapter 4 will instruct you on setting up a test plan to ensure that you know exactly what you're going to do as part of the assessment. Finally, Chapter 5 will take you through the details of testing with real users. A handy checklist with page number references back to the relevant sections can be found in Appendix A at the end of the book. Make copies of it and use it to ensure that you perform all the necessary steps as you do your usability assessments.

To help you better understand each of the steps, I'll describe the process in the context of a usability assessment performed on a product I worked on for several years. This contextual presentation should aid you in several ways. First, it will clearly demonstrate that these are the steps that are *actually* performed in a real-world usability assessment, and not an academic perspective of what *should* happen. Second, it will demonstrate that, as with any practical endeavor, there is some messiness involved in the application of the various methods, but that this messiness does not invalidate the benefits of performing usability assessment.

The product I'll be describing is the DSL self-installation kit that was developed and sold by AT&T (then known as SBC Communications) in the late 1990s and early 2000s. DSL was a type of Internet access technology that allowed broadband Internet connections to be carried over a normal telephone wire. Today, broadband Internet access is nearly ubiquitous among consumers. However, in the late '90s, broadband was still considered a technologically complex product used by a fairly sophisticated set of users.

At that time, almost all of the DSL installations were being performed by a company technician dispatched to the customer's home. A small number of installations were initiated as self-installations by customers. But even the majority of those required a technician to be sent to the user's home to complete a successful installation because the kit was essentially made up of the components that a technician would use, with little regard for how it all went together. Clearly, this wasn't a sustainable business model, as DSL was expected to become a ubiquitous means of delivering high-speed Internet, with the ultimate goal of installing millions of DSL lines each year.

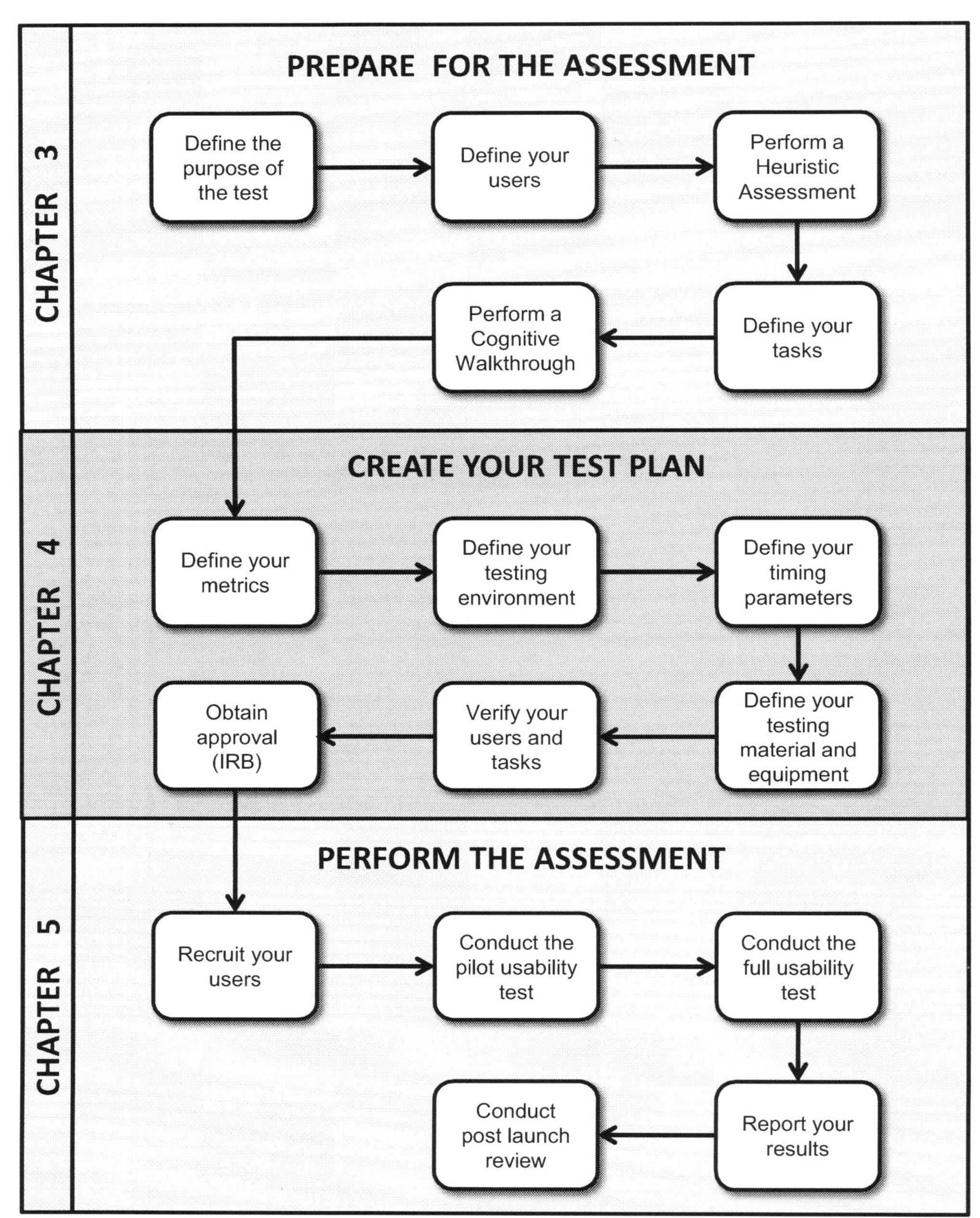

Figure 6: Steps for preparing for and conducting a rigorous usability assessment. Note the related chapters for each of the major steps.

The problem was generally acknowledged as being critically important to the company, but there were some who felt that self-installation was a pipe dream and that no amount of usability efforts would ever make the kit easy enough for average consumers to install themselves. There was recognition that even if the first version of the kit was usable, it would be used by early adopters with a fair amount of expertise, and that as broadband became more common, users with no technical expertise at all would be ordering and installing DSL. Clearly, the usability efforts on the kit would have to be rigorous to make DSL self-installation successful.

3B) Define Your Purpose: Formative Versus Summative Assessments

The first thing we did after deciding that we wanted to make the DSL self-installation kit work for all of our customers was to decide why we were doing the assessment. Was the goal to provide input to an ongoing development process? Or, rather, was the goal to establish a usability benchmark against which other products or other versions of the product could be compared?

The terms *formative assessment* and *summative assessment* describe these two different goals. Interestingly, these terms have their origins in the education field, where the assessments were focused on student learning. There, formative testing was described as low stakes and diagnostic in nature, to help guide the students toward their learning objectives, whereas summative testing was used to describe a high-stakes end-of-semester evaluation such as a final exam (Scriven, 1967). The usability community has adopted these terms to describe the activities that take place in two different forms of usability assessment.

Table 1 shows the characteristics that are typical for formative and summative usability assessments and the situations in which they are most commonly used.

Table 1: Characteristics of Formative and Summative Usability Assessments

Formative Assessment	Summative Assessment
Characteristics	
• Typically done as a series of tests	• Done as a single test
• Smaller number of users per test	• Enough users for statistical analysis
• Focus on finding usability deficiencies	• Focus on metrics of usability
• Less experimental control is allowed	• High experimental control
Common Uses	
• Part of user-centered design (UCD) development process	• Benchmark or capstone to a series of formative tests
• Examination of a specific usability issue	• Determining if a product meets minimum usability requirements for sale or procurement
• Assessment of multiple parts of a design	• Predicting user support resources needed for first launch products
• Identifying usability issues for development	

In a *formative assessment,* the goal is to gather information about a product's usability deficiencies to assist the development team. The focus of the testing is not on the acquisition of raw performance data, such as time on task or subjective usability scores; rather, the intent is to identify and understand usability issues that lead to users' inability to perform a given task. Because of this altered focus, formative assessments are often run in an iterative fashion, in a test-fix-retest methodology.

Early stages of formative testing are often done with users who may not match the exact user demographic that has been specified for the final product; this is done to save time and money. Formative methods are often considered an integral part of the user-centered design (UCD) process (Vredenburg, Isensee, & Righi, 2001; Vredenburg, Mao, Smith, & Carey, 2002) and are key to an agile, or rapid, product development process.

Over the course of the DSL self-install development effort, we conducted dozens of formative assessments. Some were run to see how well specific new design elements, such as a new instruction manual in the kit, would perform. Others were performed to determine if specific deficiencies in the design that had been identified previously (for instance, the confusability of the Ethernet and telephone cables) had been satisfactorily addressed before we made further improvements. That said, we didn't start with a formative test but, instead, performed our first usability assessment as a summative test.

For the DSL kit, we started with a summative assessment so we could establish a set of usability benchmarks. It was important for us to understand how well the product was working from a usability standpoint in its current form. The goal of a summative evaluation is not to identify specific usability deficiencies but, rather, to determine the baseline performance parameters on a given set of tasks. Summative testing answers the question, "How usable is the product right now?" and is frequently performed as a capstone to a series of formative tests to verify that the previously identified usability deficiencies have been fixed.

Summative tests are often performed on a product that the company hasn't developed itself so it can have a better idea of the product's usability. It might be that the company wants to resell that product and is concerned about maintaining minimum usability standards. Or perhaps the company is thinking about purchasing the product and wants to ensure that employees will be getting a sufficiently usable product. Summative testing is also a useful tool when trying to select from one of a number of products that the company may be considering in its acquisitions process.

One interesting use of summative testing is to help anticipate customer support requirements at the time of product launch. By understanding the number and severity of usability issues in a product, a company can make estimates of the impact on technical support resources before the product is launched.

You may not want to wait until the final version of the product before performing a summative assessment. In fact, like many other service-oriented companies, AT&T couldn't wait until every single technical and usability design issue had been fully resolved before shipping a particular version of the DSL self-installation kit. Instead, there were fixed development cycles, and the best kit available at the end of each cycle was considered production-ready. As a result, we performed a summative usability assessment prior to the launch of each version of the DSL self-installation kit to help us determine what kind of support issues might be encountered at launch. These assessments also provided benchmarks against which we could measure future versions of the kit.

It might seem intuitive that the usability would increase monotonically as later versions of the kit were released, but this was not necessarily the case. Newly applied technologies, changing business partnerships, and major revisions to installation software often caused global usability measures to fluctuate from version to version. However, by the time the kit reached its final, stable version, successful self-installation rates had risen substantially, from less than 10% to over 90%. Further, the need to dispatch a technician to the customer's home was almost eliminated (Kortum, Grier, & Sullivan, 2009).

Although these descriptions might seem to suggest that formative assessments are somewhat less rigorous than summative assessments, this is simply not the case.

The goals are different, so the focus, setup, and execution of the assessments are different.

In general, if you are interested in finding and fixing usability deficiencies in your product, use a series of formative assessments. If you want to establish a set of usability metrics to benchmark your (or someone else's) product for any number of reasons, a summative usability assessment is what you want to perform.

In the end, the repeated application of formative and summative usability assessments for the DSL self-installation kit proved to be financially beneficial to the company and to the customers who used the product.

3C) Define Your Users

Identifying the set of users who will use your product is one of the most important activities you'll perform as part of the usability assessment. No matter how rigorous your tests, how careful your planning, or how detailed your statistical analysis, if you have selected the wrong users, the output of your usability assessment will not provide the benefits you expect. In the worst-case scenario, picking the wrong users can be harmful to the company's bottom line and the reputation of your assessment group.

Let's say, for example, that in our first usability assessment of the DSL self-installation kit we had chosen colleagues from the electrical engineering laboratory to perform our first summative test. Given their intimate knowledge of the technology, it's likely that they would have been successful in completing the self-installation task. This, in turn, would have caused the human factors group to give the kit its stamp of approval, telling the company that the kit was in excellent shape and that exceptionally low levels of tech support would be required upon launch. Of course, the reality would have been much different.

So, who should we use for our DSL participants? Selecting the correct users is typically a simple matter of matching the skills, knowledge, and experience of your test participants to those of your expected consumer. In the case of DSL, however, this group of end users was a moving target. Early versions of the kit were expected to be used by people who were reasonably sophisticated technology users. We realized that as the benefits of broadband Internet access became more apparent to the general public, this user demographic would become less and less technically sophisticated as time went on. This meant that what was usable in version 1 of the kit was probably not usable in version 5, and we had to be careful to select users for each test that reflected this fact.

That said, the first versions of the kit were tested on users who already used broadband or who had self-identified as early technology adopters and were interested in getting broadband Internet. Tests performed in the later years of the kit used participants from a much broader demographic and essentially included anyone who was a telephone customer.

It's tempting to broadly define your user profile. Although this makes finding users to participate in your study significantly easier (a topic covered in section 5A), it can make interpreting the results of your usability assessment significantly more difficult. If we had recruited from the general population for the first DSL assessments, it's likely that some of those users wouldn't even know what the Internet was. (Remember, this was when the Internet was still pretty new!) Most of the participants certainly wouldn't have had any idea what broadband was. That would have made interpreting their failures very difficult. If the product and its use didn't make any sense to users, how could their performance with the product provide any insight for us?

In some cases, a broad profile is warranted. In the early 1990s, telephones and telephone services were expected to be used by everyone because landline phone penetration was greater than 96%. Recruitment of participants in usability assessments for these kinds of telephony services had to match this very broad demographic. However, in most cases, your user demographic will be much narrower.

In the case of DSL self-installation, we considered the two classic important questions when selecting our participants: (a) who will actually use the product, and (b) what will they use it for? It's important to differentiate between the user and the purchaser of the product, because the person who purchases the product (say, Mom or Dad) is often not the one who will use it (Junior), and you want the characteristics of the actual user. The distinction of what users will do with the product is also important because it might lead to the selection of different user groups, if different groups of users will use the product for substantially different tasks.

Gather information about your projected users from engineering specification documents, marketing profiles, the product manager, and the research director. Common global criteria, often found in marketing profiles, include age, gender, race, and socioeconomic status. AT&T's marketing folks were able to supply us with a detailed demographic of who was going to buy DSL.

As can be seen in Table 2, these criteria are often used to capture a number of underlying constructs. When possible, use those underlying constructs in your specification of your user, since that is the actual attribute that you are seeking. Again, make sure that you are certain that the user profile you establish matches the consumer who will actually purchase and use your product.

Expertise is an especially important selection criterion, but it can be difficult to quantify accurately. For example, using a wireless connection eight hours a day is substantially different from being able to configure an 802.11 b/g/n wireless router. A careless specification of expertise such as "Are you familiar with wireless routers?" might lead to the selection of the wrong kind of users, given that the question is open to interpretation. If I needed users who had expertise in configuring routers, this question might also capture people who are perfectly familiar with *connecting* to a router but who have no other expertise.. This distinction between true experts and mere users has been described as the difference between digitally savvy and technologically savvy customers (Change the Equation, 2015; Hargittai, 2010). Digitally savvy users may use different forms of technology extensively but have little understanding of the underlying operation or setup. Conversely, technologically savvy users more fully understand the technological underpinnings of the technology.

Hours of use of a particular technology may seem like a reasonable measure on which to screen users, but it might not get at the skills that are fundamental for your particular user set. Determine what characteristics the actual users of the product will have, and then define your test users carefully to ensure that they match those characteristics.

User characteristics for the first DSL self-installation kits reflected a typical early-adopter profile: young to middle aged, typically male, with significant technical expertise and enough disposable income to afford state-of-the-art technology. By the end of the usability assessments, this profile had changed dramatically and more closely matched the broader criteria used in earlier telephone services: a wide age range of both genders, a broader socioeconomic range to reflect the declining cost of the technology, and a redefinition of expertise to include occasional computer and Internet use.

Table 2: Common Selection Criteria and Their Underlying Constructs

Selection Criteria	Underlying Constructs
Age	• Willingness to adopt new technology • Understanding of technology • Cognitive deficits • Perceptual deficits • Physical deficits
Gender	• Physical differences (e.g., strength, height) • Differences in problem-solving approaches • Different use patterns • Perceptual deficit
Socioeconomic status/Income	• Technology ownership • Technology adoption • Expertise
Race	• Socioeconomic status • Technology use
Expertise	• Raw intelligence (IQ) • Frequency of use for a given technology • Duration of use for given technology • Understanding of the underlying technology • Ability to troubleshoot

Some usability researchers believe that selecting users with specific characteristics is less important than I have described here. They suggest that you should start by testing somebody, *anybody*, if you can't get your exact target demographic (e.g., Goodman, Kuniavsky, & Moed, 2012). In some cases, some data is better than no data, but it is often the case that some data can actually be worse than no data if it leads you to make the wrong conclusions. Define and select your users carefully and you will have substantially higher confidence in your results.

3D) Perform a Preliminary Assessment I: Heuristic Assessment

Performing a heuristic evaluation is one of the first steps in undertaking a complete usability assessment, and was certainly the first thing we did with the DSL self-installation kit. A **heuristic evaluation** is a method in which an evaluator assesses a product or service and determines its compliance to a set of known usability principles. These principles are called *heuristics.* First described by Nielsen and Mölich (1990), and later refined by Nielsen (1992a, 1994a), a heuristic evaluation is one of the most common usability methods employed because it's highly cost-effective, reasonably easy to learn and use, and can be done relatively quickly.

Shouldn't I develop and use "personas" to characterize my user population?

No! Although personas are a wonderful shorthand method to describe specific demographics of your user population and have been used to good effect (Miaskiewicz & Kozar, 2011), their downsides outweigh their benefits (Chapman & Milham, 2006; Portigal, 2008). Teams often create a host of personas to represent their user population and describe these prototypical users in great detail (e.g., "John likes to play pick-up basketball on the weekends at the local gym"). Many teams even attach pictures to these profiles to complete the persona and refer to the personas as they work through the usability assessments.

The problem with personas is that they tend to take on a life of their own. Their function as a prototypical description of user characteristics gets forgotten, and the usability professional starts to fill gaps in the persona with stereotypical elements for people "like that." Worse, personas often begin to look suspiciously like the people who developed them, and the persona becomes a proxy for "ME." As I have described earlier, you are never the user, and using that persona as a guide can only lead to trouble.

Can't I just use other people on the project as usability test participants?

No! Other people on the project, particularly the programmers and developers, have special knowledge of your product or system. They understand how it works. They probably had a hand in developing specific features and functions in the interface. This specialized knowledge means that they are not representative of the people who will eventually use your system. If your product or service is only going to be used by your programmers and developers, this is certainly the appropriate test group. However, if your product or service is going to be used by the general public, it is very important that you test with users who represent a broader demographic.

What about using programmers and developers from other projects in your company? If this is the only resource available to you for participants, this can be a workable solution. However, the same admonition holds. These users have specialized knowledge, and the results that you obtain will likely put your product in a much more favorable light than it would have been had you tested with more representative users. This might sound like a good outcome, but remember that the goal of usability testing is to uncover and fix usability issues, not to put your product in a favorable light.

Table 3 shows the nine heuristics identified by Nielsen and Mölich (1990). These heuristics are general enough to be useful for a wide range of products, services, and systems. That said, it's not uncommon for practitioners to add or modify specific heuristics to suit their unique situation, although some prominent researchers don't condone this practice (Sauro, 2015).

With adherence to the underlying principles of constructing a good heuristic, we chose to add some DSL-specific heuristics, which are shown in Table 3. When adding or modifying heuristics, it's often helpful to understand what heuristics are not. Most important, a heuristic is *not* a checklist or set of specific requirements. When evaluating a product against a requirements list, an untrained individual can simply go through and determine if the product meets that specific requirement. For example, a requirement might state, "There shall be a keyboard shortcut for printing." An evaluator could simply determine that there was a keyboard shortcut for printing a document and mark that item as complete. The general heuristic for this would be "flexibility and efficiency of use" and "shortcuts should be available," but the absence of a specific feature (like the keyboard shortcut) would simply be identified as a deficiency in ease-of-use along with any other items that failed to meet the heuristics.

In the same vein, heuristics are *not* standards. Standards (e.g., ISO, ANSI, SAE) are often used as requirements to ensure that a product complies with a known set of design elements, whereas heuristics are significantly more general in their application.

Table 3: General-Purpose Heuristics Proposed by Nielsen and Mölich (1990), and Some Specific to the DSL Self-Installation Kit

General Purpose

- Employ simple and natural dialogue
- Speak the user's language
- Minimize the user's memory load
- The interface should have consistency
- Sufficient feedback should be provided
- Clearly marked exits should exist
- Shortcuts should be available
- Precise and constructive error messages should be used
- Help and documentation should be available

Specific to DSL Self-Installation

- Minimize physical alterations to equipment at the customer's home
- Sequence of events and operation should be evident

The evaluation is performed by an expert, who examines all of the important aspects of the interface. The resulting output of the heuristic evaluation is a list of usability deficiencies that can be mapped back to each of the heuristics.

Looking at some of the outputs from the DSL self-installation kit heuristic evaluation should help to clarify how the method is used. One of the earliest versions of the kit was essentially a collection of parts used by the technician to do a professional installation. Each of the components (the modem, Ethernet card, signal-splitting device, Internet service provider software, etc.) had its own installation manual and installation CD. The installation also required the user to rewire the telephone interface box on the outside of the home to get the Internet signal inside.

Some of the heuristics resulting from an assessment of this kit included the items shown in Table 4. (Table 4 is only a partial listing of the output of the heuristic evaluation. The actual list would have included hundreds of items across all of the heuristics.)

It's important to always perform a heuristic evaluation of the product before you do your usability assessment with real users, because it helps prepare you for user testing. First, it helps you gain familiarity with the product and understand the kinds of usability deficiencies from which it suffers. The heuristic evaluation also provides an important preview of error categories, which can be beneficial during actual user testing. Finally, it can aid in the definition of tasks that will be used during the cognitive walk-through (section 3F) and actual user testing.

Table 4: Example Outputs for a Heuristic Evaluation of One of the Earliest DSL Self-Installation Kits

Heuristic	Usability Deficiency That Violates the Heuristic
Speak the user's language	• Many of the instructions refer to the NID, and NID (network interface device) is terminology specifically used by telecommunications professionals.
The interface should have consistency	• The customer must make significant alterations to the telephone interface box, and these modifications are not the same at every customer's home. • Each of the installation CDs has a completely different interface.
Sufficient feedback should be provided	• The customer is told to proceed "when the Internet connection is complete," but there is no clear indication of this state in the software. • The customer cannot easily determine if the signal coming to the modem is the Internet signal or the telephone signal.
Minimize the user's memory load	• A lengthy code must be memorized or written down and then transferred to one of the input screens.
Sequence of events and operation should be evident	• It is unclear which of the components should be installed first, as several of the manuals say "read me first."
Minimize physical alterations to equipment at the customer's home	• The customer must make significant alterations to the telephone interface box. • The customer must open the computer to install a network interface card to get the modem to be functional.

Because the heuristic evaluation method described here is used only in preparation for actual user testing, a single evaluator (you) can perform the evaluation. If you're using heuristic evaluation as the sole assessment method, five or more expert evaluators must be employed to gain the maximum benefit (Nielsen, 1994a).

3E) Define Your Tasks

As noted earlier, defining your users is one of the most important things you will do as part of your usability assessment. That said, defining your *tasks* is a close second. The number and type of tasks that you define for your users to perform will largely determine the kinds of usability deficiencies that are identified. The more functionality a product has, the more difficult it can be to define the tasks.

Task definition for the DSL self-installation kit was straightforward. Although the kit had many pieces of hardware and software, it had but a single function: to enable a user to install high-speed Internet in his or her home. When users came to the laboratory, we handed them a sealed box as if UPS had just delivered the kit and gave them a single instruction: "Use this kit to get your high-speed Internet service working on this computer, and tell us when you are finished." This simple, top-level task encompassed all the necessary subtasks that a user would need to perform to be successful with this particular product.

Heuristic evaluations are cheap and easy. Can't I just do that?

No! Some research indicates that expert reviews can provide the same quality data as user testing (Molich & Dumas, 2008), but the results you obtain will not be nearly as accurate or informative as the results obtained if you actually put your product in front of users.

Research has demonstrated that heuristics alone identify only a half to a third of the issues that are found in an actual user test (Desurvire, 1994; Hvannberg, Law, & Lérusdóttir 2007; Sauro, 2012). That's not to say that discount usability techniques like heuristic evaluations are not effective on their own. Many organizations use them (often exclusively), because they are extremely cost-effective and relatively quick and easy to implement. However, users can be extremely unpredictable, and even the best usability expert will often miss user behaviors that surface during a usability test (Jeffries & Desurvire, 1992).

The best benchmark for the usability assessment of a product or service is a usability test in which representative users exercise all of the important features and functions of a product, as only a usability test represents "real" data in the minds of many developers (Kantner & Rosenbaum, 1997). If you want to have the highest confidence that you have identified the most serious and important usability issues, you should always test with real users and gather evidence-based data.

On the other end of the spectrum, imagine trying to develop a task list for a piece of spreadsheet software like Microsoft Excel. The program contains hundreds of functions that specific users would consider key elements of the interface, and testing every possible way the program could be used would be nearly impossible.

There are several ways to approach the problem of "too many tasks." One of the best ways for general consumer products is to identify those features and functions that will be used by the largest number of users, and/or those functions with the highest frequency of use. If, for example, you are testing Amazon.com, purchasing an item from the site would certainly be one of the major tasks that would need to be included. The advantage of this approach is that it optimizes the use of your usability assessment resources and maximizes the likelihood that you will identify the most impactful usability problems in the product.

You must be cautious when examining frequency data for your task selection. Certain critical tasks (such as signing in to a site) may occur only once, but their frequency does not provide an accurate representation of their importance.

Another method of identifying tasks for inclusion is determining their criticality. Identifying usability deficiencies for those tasks that could cause the most harm (rather than those that derive the greatest benefit) can be reasonable for certain kinds of systems in which usability failures can have significant adverse consequences. In certain types of mission-critical interfaces, such as equipment used in human spaceflight operations, even selecting the most critical tasks is not sufficient; it might be necessary to perform comprehensive usability assessments when an exhaustive list of all the features and functions of that equipment is evaluated.

If time and budget permit, another good approach is to evaluate a broader range of features and functions, but doing so in a series of smaller formative tests. In this way, more of the product can be evaluated, but the task is divided into reasonably sized pieces that can be readily tested.

What about impossible tasks?

Sometimes you want to know if users know when they *can't* do the task. In this case, you might ask them to perform a task that can't be done with your interface. This is especially important for certain kinds of interfaces, such as those used in information search or in large e-commerce sites. Here, it's important for users to know when to quit. If the information (or a specific product on an e-commerce site) isn't available and the user can accurately determine that it isn't available, the interaction can be considered successful. The key is to give proper instruction to users. This is an example of a good instruction:

"We're going to ask you to find a number of pieces of information. This information may or may not be on the site. If you think the information isn't on the site, tell the experimenter."

Instructions like this will help to ensure that users will work on the task for a realistic amount of time, just as they would at home, before they give up.

The tasks you define here will be used by *you* during the cognitive walk-through (described in section 3F) and by *your users* during the actual usability test (described in Chapter 5), so it is imperative that the tasks are representative and carefully selected and described.

3F) Perform a Preliminary Assessment II: Cognitive Walk-Through

Performing a cognitive walk-through is the next step in getting ready for testing with users. Like a heuristic evaluation, the cognitive walk-through is performed by an expert evaluator rather than actual users. What differentiates the cognitive walk-through from the heuristic evaluation is that the cognitive walk-through is *task*-based. Using the tasks developed in section 3E, the evaluators walk through the task, step by step, to determine where users may have difficulty.

The theory behind cognitive walk-throughs was first described by Polson, Lewis, Rieman, and Wharton (1992), and later detailed in a form more suitable for practitioner use by Wharton, Rieman, Lewis, and Polson (1994). In a cognitive walk-through, the evaluator assumes the role of the user and then, in his or her mind, walks through the process of performing the task. As they walk through the task, they answer four questions for each step:

1. Will users try to do what needs to be done to perform the task?
2. Will users be able to see that the interface provides a way to perform an action to move toward completion of the task?
3. Once users find the correct action on the interface, will they know that it is the right one for what they trying to do?

4. After the action has been taken, will users understand the feedback they get and understand they are making progress toward their goal?

In practice, these four questions can be more cumbersome than is necessary. Spencer (2000) developed a reduced set of questions that are more constrained and speed the evaluation process. This reduced set contains only two questions:

1. Will users know what to do at this step?
2. If the users do the right thing, will they *know* that they did the right thing?

Even if it is determined that the user is going to have significant difficulty in determining what to do, the evaluator should, after making note of the deficiency, assume that the user has performed the correct action and then assess Spencer's second question of whether the user will know that she did the right thing. In this way, the analysis can proceed step by step even if the interface is highly deficient.

As can be seen, the method focuses on the learnability of an interface. It has value above and beyond the heuristic because it identifies specific weaknesses associated with the sequence of events that must take place for a task to be completed successfully. It is worth noting that because the method describes the way in which the user is unable to complete the task, it provides excellent input into the design process to determine how the problem might best be resolved.

In the case of the DSL self-installation kit, performing a cognitive walk-through on the single global task of getting high-speed Internet service to work would result in a lengthy analysis comprising each of the substeps necessary to perform an installation. For exceptionally complex systems, this might seem to make the cognitive walk-through unwieldy. In such cases, subtasks could be evaluated independently, provided that an assessment was made of how users would know to move from subtask to subtask (as determined with a separate cognitive walk-through). In this way, a cognitive walk-through could be performed more efficiently, as multiple evaluators could be employed for the different subtasks.

Table 5 shows a series of tasks performed as part of the DSL self-installation and the responses to the two questions posed in Spencer's streamlined method. For each task, the answers to question 1 are determined by evaluating the interface and determining which interface elements provide action options for the user. For example, in task B (getting instructions), the user is presented with a box containing many instruction manuals. By thinking about how the user might choose among these manuals, the evaluator can easily make the determination that a user is going to have to examine the different manuals and make a decision about which one is the correct one. Because there are multiple options, this element of the interface clearly needed more work to make it easier to use.

The second question we ask, "Will the user know they did the right thing?" is evaluated by the feedback the user gets *if* he or she chooses the correct path. In this case, if the user chooses the right manual, it says "READ ME FIRST," an excellent clue that he's on the right path and has chosen correctly. This evaluation is completed for each of the specific subtasks a user must perform to complete a given task.

Table 5: Cognitive Walk-Through Outputs for Select DSL Self-Installation Tasks, Using Spencer's Reduced Question Set

Task	Question 1: Will the user know what to do at this step?	Question 2: Will the user know he/she did the right thing?
A. Start the installation after receiving the kit from the postal service	• The user is expecting the kit, and opening a box from a vendor is a standard activity upon getting a box in the mail. • The top of the box says "Open the box to start your DSL self-installation" in prominent text.	• When the box is opened, the contents of the kit are visible.
B. Get the instructions to start the installation	• There are multiple sets of instructions and marketing materials in the box. • The cover designs on each of the instructions are similar.	• The correct manual is clearly and conspicuously marked with a label that says "READ ME FIRST."
C. Connect the DSL modem to the DSL signal at the wall plate	• Multiple jacks on the back of the modem will accept the phone connector. The modem operates only if the correct jack is used. • The jacks are not color-coded or labeled, so it is not clear which jack is correct. • In newer homes it may be unclear whether the Ethernet cable or the phone cord should be used to connect the wall plate to the modem.	• The SYNC light will begin to blink red upon connection and eventually turn green. It is unclear whether users will understand that SYNC means the modem is connected.
D. Install the DSL software on the computer	• There's only one CD in the kit. • Instructions clearly tell the user when and where to insert the CD. • There is nothing to prevent the user from inserting the CD before the appropriate time.	• Once inserted, the CD auto runs and launches the installation program.

Note: Task A requires no further action, whereas tasks B, C, and D would require further design effort.

The table shows that in some cases the interface performs well and the user is able to move through the task with a minimum of effort (A). In these cases, no additional design efforts from a usability standpoint should be required. In other cases (B,

C, D), the user is likely to experience difficulty in completing the task, and this failure should be noted for further design review. The cognitive walk-through also helps guide you in determining which tasks might require increased scrutiny during the testing of human participants.

Do I always have to do a heuristic evaluation and cognitive walk-through before testing with users?

You could skip these steps, but the output from your usability assessment won't be nearly as good. Performing these two analyses helps you prepare for the usability test. It helps you to better understand potential pitfalls that your users will encounter, and it helps you determine the kinds of tasks that users will be evaluating during your test.

Some experienced usability professionals do not take the time to formally go through a heuristic evaluation or a cognitive walk-through prior to usability testing to create tasks and gain a better understanding of the kinds of errors that users may be making. However, the steps they take to prepare for a usability test often encompass many of the same procedures used in heuristic evaluations and cognitive walk-throughs. Formally performing these evaluations simply helps to ensure that you will be running the most effective usability assessment possible.

3G) Suggested Reading

- "Heuristic evaluation" by Jacob Nielsen. In J. Nielsen and R. Mack (Eds.), *Usability Inspection Methods* (pp. 25–62). John Wiley & Sons, 1994.
- "The cognitive walkthrough method: A practitioner's guide" by Cathleen Wharton, John Rieman, Clayton Lewis, and Peter Polson. In J. Nielsen and R. Mack (Eds.), *Usability Inspection Methods* (pp. 105–140). John Wiley & Sons, 1994.
- "The streamlined cognitive walkthrough method: Working around social constraints encountered in a software development company" by Rick Spencer. In *Proceedings of the SIGCHI Conference on Human Factors in Computing Systems* (pp. 353–359). Association for Computing Machinery, 2000.

CHAPTER 4: CREATE YOUR TEST PLAN

Now that you've completed your heuristic evaluation, defined who your users are, and conducted a series of cognitive walk-throughs using the tasks you've developed, you're ready to use that information to complete your test plan. The test plan will serve as an outline for all the actions that you will undertake to run your usability assessment with real users. It helps to ensure that you don't miss any of the important details of the test, and, more important, it forces you to consider all of the elements of your testing *before* you begin.

Not only does the test plan contain information about your users and tasks from the previous evaluations you have just completed, it also describes which measures you will use, what your testing environment will look like, what equipment you will need, and how the test will actually be run. Each of these elements will be covered in turn in the following sections. Table 8, at the end of this chapter, contains a list of all the major things you'll need to define as part of your test plan.

Experienced usability professionals may tell you that they don't go through the formal step of constructing a test plan in writing. Although foregoing the written plan might seem like a time-saving strategy, the test plan serves two other important functions. First, it documents what you're going to do in a way that can be shared with other people, such as your manager and others on the product design team. Second, it serves as an excellent basis for the final report that will result once the usability assessment is complete.

In the case of the DSL self-installation assessments, the written test plan was important for ensuring that the consistency of the evaluations was maintained across the multiple tests, increasing the reliability of the results and the team's confidence in the procedure over time.

4A) Define Your Metrics

Whether you're running a small formative test or a larger summative assessment, you'll need to define exactly which metrics you will use to measure the usability of your product, service, or system. Although many measures can be used, most usability professionals rely on a small, fairly standard set of metrics to measure usability.

4Ai) ISO 9241–11

The three most common and important metrics employed in usability assessments are

1. Effectiveness
2. Efficiency
3. Satisfaction

These measures have been agreed upon by a group of experts who have created a formal document, called a *standard,* that describes what is considered the best practice in a given discipline. These standards can then be used by other professionals to help create a common language and practice of use. Further, these standards can be used as shorthand when specifying what needs to be done for a given activity.

In the case of usability, the International Organization for Standardization (ISO) has identified these three metrics in the standard ISO 9241-11:1998, *Ergonomic requirements for office work with visual display terminals (VDTs)–Part 11:*

Guidance on usability as the most appropriate ones to measure how well users can achieve their goals when using a given product.

Effectiveness. Effectiveness is the ability to perform a task. It is typically measured using success rates and/or counting the number of errors that a user commits in trying to perform the task. At first glance, success on task seems like a trivial thing to measure; you either complete the task or you don't. Some tasks lend themselves very well to this simplistic reading of the measure. The DSL self-installation process was considered to be successful if the user could surf the Internet with his or her high-speed connection after the installation was complete. This black-and-white definition of success or failure greatly facilitates the reporting of results in the form of a simple percentage of participants who were able to perform a specific task.

Unfortunately, this binary definition of success or failure may not convey all of the important information about the user's actions. In DSL self-installation, it's possible to connect the modem in such a way that you can still surf the Internet, but only at greatly reduced speeds. Is that a success? This example highlights the importance of specifying success with a great degree of precision. In the case of DSL, the success metric could be specified as surfing the Internet at speeds above 1.5 Mb per second. In this case, a user who is able to surf the Internet but only at reduced speeds would be classified as a failure.

It has been suggested that success is the simplest and most important metric (Nielsen, 2001), because if a user can't do the task, nothing else really matters. Although this may be true in summative assessments, it certainly isn't true in formative assessments, in which a greater understanding of the reasons for failure is required.

Because of the ambiguity that sometimes surrounds what constitutes success and the need for a better understanding of why people don't succeed, we need to think about error rates, which is another way to measure effectiveness. Defining what constitutes an error can vary tremendously from product to product. Whole taxonomies have been built to specify error (Reason, 1990), but many of them are too unwieldy for practical use.

In general, you're trying to capture three kinds of errors. The first is an error in which a user makes a conscious choice do something, but it is the wrong choice (an *error of commission*). For example, a user would have committed this kind of error if, after examining the back of the modem and the DSL signal cable, decided to plug the DSL signal cable into the Ethernet port instead of the RJ-14 port.

The second kind of error is one in which the user simply fails to perform a necessary task (an *error of omission*). A user who didn't plug the DSL signal cable into the modem would have committed this kind of error. The third kind of error is one in which a user might do the right things but do them in the wrong order (a *sequence error*). For example, a user trying to install the connection software for DSL before making the physical connections to the modem would be making a sequence error.

In a usability assessment, the goal is simply to *capture* the errors; classifying them is somewhat less important. The metric can be reported as the number of errors committed by the user, or the error rate as a function of the total number of errors that could be committed. The specific errors should also be reported to help the design team understand why people didn't succeed.

Remember, it's not necessary for an adverse consequence to occur for an error to have occurred. Even if the user makes an error but recovers from it, the error should be captured and noted as part of the usability assessment to aid the design team in understanding how the product operates.

Efficiency. Efficiency is the effort required by a user to complete a task. One common measure of efficiency is the time it takes a user to complete a task. Although the concept of time on task is simple, it can be more difficult to measure reliably in the field. See section 4C for a more detailed discussion of setting your timing parameters appropriately.

Total time on task is important, but it's also common to want to have efficiency measures for subtasks within the main task. In this case, you would develop timing landmarks for the global task and timing landmarks for each of the subtasks. Be sure to think about the kinds of activities that will occur during the completion of any given task, and then determine if you're going to include these activities in the total time. For example, are you going to count the time that the user spends reading the manual? What about the time that the user spends working with the help desk, either by phone or by chat? Generally, you should count all of the time that it takes users to complete the task, regardless of what that activity was, discounting only personal interruptions such as a restroom break or personal phone call.

As long as you log all the users' activities and their associated times, you can always go back and remove items that you've determined aren't part of the actual task. Once the test is complete, however, it's impossible to add these back in. For example, during the DSL testing, one of our users decided to call a friend for assistance. The friend was busy and said that he would call the participant right back. The user then sat down and read a book until his friend called back. In this case, we recorded the entire completion time but ended up taking out the time from the end of the initial call to the start of the call when his friend called back.

One common question that arises in the measurement of efficiency is how long you let the user struggle before you stop the task. Certainly, if a user self-identifies that he or she wants to give up, mark that task as a failure, note that the user gave up, and stop the timing. Some users just won't give up, however, so before the test begins, set an upper time limit for a given task. At that upper time limit, stop the task and note it as a failure-to-complete within the given time.

Setting the upper time boundary can be tricky. It can be anywhere from 3 to 100 times the length of the average success time, depending on the length of the task. For example, if average success time on a task is 1 minute, 3 minutes might not be sufficient to capture the variance in completion times, and so a longer upper boundary of 10 minutes might be reasonable. On the other hand, if average completion time is 1.5 hours, 5 hours as an upper bound might be excessive. In these kinds of cases, the limits of your user's attention and stamina might guide the upper limit.

Besides time, a number of other common measures of efficiency include counts of specific discrete actions performed by the user, or measures of other continuous performance variables, as shown in Table 6.

Satisfaction. Satisfaction is users' subjective impression of how well a product or service meets their personal expectations and can include hedonistic qualities such as desirability and pleasurable use. Some purists believe that satisfaction isn't really an important part of usability and that if a user is successful, it doesn't matter whether or not he or she is satisfied. If the user has no choice in the selection or use of the system (for instance, a military missile system or a phone issued to you by your employer), one might argue that satisfaction *isn't* a variable of importance. However, even in such cases, decreased satisfaction can lead to lower use, the employment of incorrect or unreliable workarounds, and even the surreptitious use of more usable systems to accomplish a goal (e.g., using an unsecured personal telephone to make top-secret calls because my secure phone is too difficult to use).

Table 6: Other Common Measures of Efficiency Besides Time

Discrete Actions
Number of mouse clicks
Number of waypoints hit
Number of specific physical motions a user makes
Number of keystrokes
Number of button presses
Number of times a user attends to a specific interface element
Number of times a user consults the installation or help manual
Number of times a user calls the help desk
Continuous Variables
Distance the mouse has traveled
Length of eye-scan paths when using the interface
Distance a user must move during operation of the interface
How lost a user is on a Web site (Gwizdka & Spence, 2007)
Global estimates of pages visited in an interface (Scharff & Kortum, 2009)

Note: Generally, higher counts on these measures signify lower efficiency.

Satisfaction can be especially important for consumer products when the user has a choice. We measured satisfaction in every one of the usability assessments we made of the DSL kit to ensure that the subjective customer experience with the kit was improving, right along with the improvements in the objective measures of effectiveness and efficiency.

The After Scenario Questionnaire (Lewis, 1991, 1995) is one good way to measure satisfaction. It has three questions, each measured on a 7-point scale, as shown in Figure 7. The satisfaction score is computed as the average rating for these three questions.

4Aii) Subjective Usability Metrics – the System Usability Scale (SUS)

One of the most common ways to measure satisfaction is with the System Usability Scale (SUS; Brooke, 1996), shown in Figure 8. It is widely used as a proxy for satisfaction, even though it actually measures subjective usability. The instrument is made up of 10 questions and yields a score from 0 to 100, which makes it easy for nonexperts to interpret.

The instrument, which was modified slightly by Bangor, Kortum, and Miller (2008) to make its language more understandable by the general user population, has well-understood psychometric properties (see also Sauro, 2011a). It is also technology-agnostic, so it can be administered across a wide range of products, services, and systems.

One of the biggest advantages of the System Usability Scale is that comparative benchmarks have been published published (Bangor et al, 2008; Kortum & Bangor, 2013; Kortum & Sorber, 2015; Kortum & Peres, 2015), enabling you to see how the usability of your product compares with that of other products. There are also data that tie SUS scores to acceptability-of-use scales that show whether or not your product has acceptable usability (Banger et al., 2009). Table 7 contains a number of comparative benchmarks from these studies, and Figure 9 shows the acceptability-for-use scale.

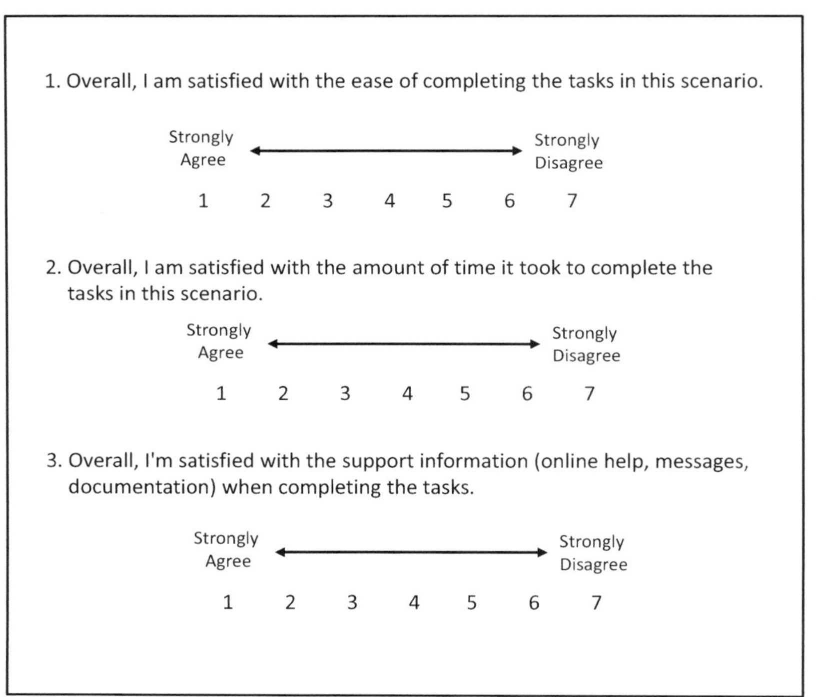

Figure 7: The After Scenario Questionnaire (Lewis, 1991, 1995) used to measure satisfaction.

It is common practice to change the word *product* to the name of your specific product or service to aid the user in making usability judgments. This change has been demonstrated to have no impact on the reliability of the survey (Sauro, 2011a).

Because the questions are worded with different polarities, scoring must be done carefully. A number of excellent automated tools on the Internet can help you do this (e.g., Jeff Sauro's calculator at www.measuringux.com/SUS_Calculation.xls), or you can calculate it by hand using the formula (1) shown here:

$$((Q1-1) + (Q3-1) + (Q5-1) + (Q7-1) + (Q9-1) + (5-Q2) + (5-Q4) + (5-Q6) + (5-Q8) + (5-Q10)) \times 2.5 \quad (1)$$

The SUS was administered after every formative and summative test of the DSL self-installation kit so we could measure progress (or lack thereof) as we continued to modify the kit. The first version of the consumer-shippable kit had an average SUS score of 62, which is in the low marginal range. Over the course of the next 11 summative usability assessments, usability ranged from 69 to 88 (Kortum, Grier, & Sullivan, 2009). As described earlier, the increase was not monotonic in nature. With the introduction of new modems, new business partnerships, and new technology, there was significant variation in the SUS scores over time.

	Strongly Disagree				Strongly Agree
1. I think that I would like to use this product frequently.	1	2	3	4	5
2. I found the product unnecessarily complex.	1	2	3	4	5
3. I thought the product was easy to use.	1	2	3	4	5
4. I think that I would need the support of a technical person to be able to use this product.	1	2	3	4	5
5. I found the various functions in the product were well integrated.	1	2	3	4	5
6. I thought there was too much inconsistency in this product.	1	2	3	4	5
7. I imagine that most people would learn to use this product very quickly.	1	2	3	4	5
8. I found the product very awkward to use.	1	2	3	4	5
9. I felt very confident using the product.	1	2	3	4	5
10. I needed to learn a lot of things before I could get going with this product.	1	2	3	4	5

Figure 8: The Modified System Usability Scale (Bangor et al., 2008)

Because of the versatility and high utility of the SUS, you should use it each time you perform a usability assessment in both formative and summative tests. It's an easy-to-use, easy-to-understand metric that will help you chart the usability progress of your product as it is being developed. It will also provide benchmark data that will enable you to compare the usability of your product against the usability of a wide range of products and services that have published SUS scores.

Likelihood to recommend (LTR) is another measure of satisfaction that has been gaining widespread acceptance in usability assessment. LTR is based in marketing theory and was described by Reicheld (2003) when he developed the Net Promoter Score to characterize customer satisfaction. It is measured simply by asking users to rate on a 1 to 10 scale, (1 = *very unlikely*, 10 = *very likely)* "How likely would you be to recommend this product to your peers/friends?"

There is a high correlation between SUS scores and LTR scores (Sauro, 2011b), given that usability plays a significant role in whether or not a consumer recommends a given product, service, or system to his or her peers (Bradner & Sauro, 2012). The LTR appears to be measuring a different facet of satisfaction than the ASQ described earlier, so if you are building commercially available products, it is an excellent measure to take and one that is especially useful when communicating results to your marketing colleagues.

Table 7: System Usability Scale Scores for a Number of Products for Comparison

Product	SUS	Product	SUS
Common Hardware		***Mobile Applications***	
• GPS navigation devices	71	• Average of top 10 iPad apps	73
• Digital video recorders (DVR)	74	• Average of top 10 iPhone apps	79
• Nintendo Wii	77	• Average of top 10 Android apps	83
• iPhone	79		
• Bank ATMs	82	***Web pages***	
• Microwave ovens	87	• Healthcare.gov	34
• Landline telephones	88	• Amazon.com	82
		• Gmail	84
		• Google Search	93
Medical Devices		***PC Software***	
• Epinephrine injector pens	65	• MS Excel	57
• Home pregnancy tests	67	• MS PowerPoint	75
• Syringes	68	• MS Word	76
• Manual blood pressure cuffs	74	• Web browsers (average)	88
• Inhalers	76		
• Temporal artery thermometers	81		

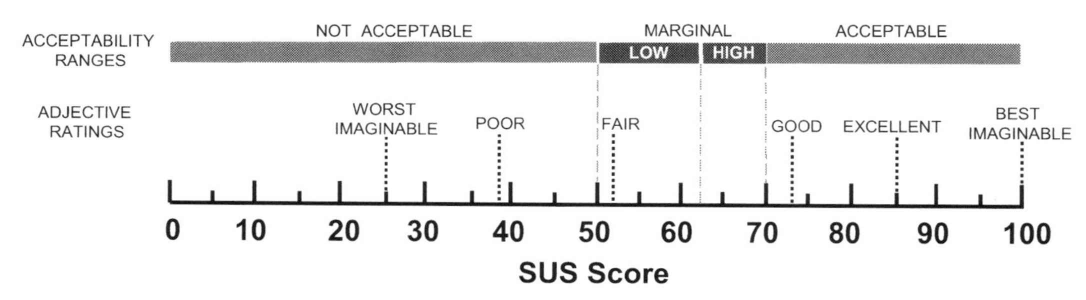

Figure 9: System Usability Scale scores mapped onto an acceptability-for-use scale, along with adjective rating that describes various SUS scores (Bangor, Kortum, & Miller, 2009).

4Aiii) Other Metrics: Learnability and Workload

Although effectiveness, efficiency, and satisfaction are the three most common metrics used to assess usability and have been codified in ISO 9241–11 as the metrics of choice, several other metrics have gained widespread use in helping to assess usability.

Learnability is often considered a key component of ease of use, and its inclusion as a usability metric seems intuitive. Nielsen (1993) included it in his early list of heuristics, suggesting that learnability is one of the basic measures for understanding how usable a system is. Lewis and Sauro (2009) even demonstrated that the

System Usability Scale may be composed of two factors – usability *and* learnability – although these results have not been widely replicated.

Other researchers have suggested that ease of use and learnability may work as opposing forces, whereby gains in one may result in losses in the other (Paymans, Lindenberg, & Neerincx, 2004). Further, no specific learnability metrics have been validated in the usability assessment community (Grossman, Fitzmaurice, & Attar, 2009), so it is recommended that the standard ISO metrics of effectiveness, efficiency, and satisfaction be used to capture the elements of usability that are associated with learnability.

Workload is another commonly collected metric. It measures the amount of effort that a user must expend to use an interface. Whereas some workload measures include the physical workload of the task, most measures focus on the mental workload aspect. Mental workload is affected by a number of factors, including the difficulty of the task, the number of concurrent tasks a user must perform, the rate of those tasks, the amount of experience or training a user has with the device, and any number of environmental factors. Generally, interfaces that have high workload are less usable than interfaces that have low workload.

There are many ways to measure workload, including performance measurements, the use of physiological measures, and subjective measures that are assessed with surveys.

Performance measures are usually assessed by employing two tasks, a primary and a secondary task. The user is given these two tasks and told to make sure the primary task is always performed. When the demands of the primary task are low, additional capacity held by the user can be used to perform the secondary task. If the demands of the primary task are high, fewer resources can be devoted to the secondary task.

One of the difficulties with performance measurements is that for simple tasks, the user never reaches his or her maximum performance level, so reserve capacity is never taxed. More important, these kinds of performance measures are typically too cumbersome for usability assessment.

Physiological measures of workload include heart rate, respiration rate, pupil diameter, eye-blink frequency, and assessments of speech patterns. Correlations between workload and these physiological measures have been established, and so the level of workload can be established by taking the appropriate measurements during a usability assessment. Physiological measures have the advantage of providing a continuous record of workload, and the user's task doesn't have to be interrupted to take a measurement.

Unfortunately, physiological correlates of workload are not always well understood, and only inferences about workload can be made. Further, reasonably large individual differences can make interpretation of physiological measures more difficult.

The recommended method of collecting workload measures for usability assessment is through the use of *subjective assessment.* Subjective measures of workload typically employ multidimensional scales and are administered as surveys. Both subjective and objective measures have value in the measurement of workload, but some studies have shown that subjective measures tend to be better (Solomon, Mikulincer, & Hobfoll, 1987; Wierwille & Conner, 1983). Subjective measures are certainly much easier to gather in real-world environments, as the user is not required to be connected to any physiological measuring devices and can focus on the primary task of interest. Workload surveys suffer from the fact that users may have preconceived notions about the level of workload for a particular task, and their responses may be biased by internal motivation, and peer or organizational pressures.

The most popular subjective workload assessment is the NASA Task Load Index (NASA-TLX), developed by Hart and Staveland (1988). Although many other subjective workload measures are available, the NASA-TLX has demonstrated that it is one of the most sensitive measures (Hill et al., 1992).

The NASA-TLX, shown below in Figure 10, measures workload along six dimensions, including physical demand. Similar to the System Usability Scale, the NASA-TLX provides a score from 0 to 100 for ease of interpretation. The original scoring method described by Hart and Staveland used a complex, time-consuming, and exhaustive pairwise assessment of the weights for each of the measures. There is mounting evidence that the pairwise comparisons do not greatly affect the score distributions and can be omitted (Bustamante & Spain, 2008; Hart, 2006). Calculation of the unweighted overall NASA-TLX scores is then computed by averaging the scores for the six dimensions (given than each dimension is scored on a 100-point scale in increments of 5). If you want to use the weighted version, scoring is most easily accomplished by utilizing one of the numerous online administration and scoring tools available.

Overall workload scores can be a good indicator for potential usability issues. If an interface has high workload, it tends to be less usable than an interface with low workload. Low rankings on the Performance subscale and high rankings on the Frustration subscale are particularly informative, as they point to difficulties that users encounter in performing their tasks in an acceptable manner.

4B) Define Your Testing Environment

Although most people think of one-way mirrors and laboratory-like testing facilities when conducting usability assessments, technology has enabled additional flexibility for the practitioner in the selection of the testing environment. There are three general kinds of testing environments that are common in usability assessment:

1. Remote environments where the user is in one location and the practitioner is in another. (You here, users there.)
2. Remote environments where both the user and the practitioner are together at the user's site. (You there, users there.)
3. Local environments where the user and the practitioner are together in a facility that has been selected specifically for usability assessment. (You here, users here.)

Each of these environments has its advantages, and multiple environments may be utilized over the course of the testing for a given product or service.

4Bi) Remote Testing: You Here, Users There

This kind of remote testing is very common when testing products and services that can be delivered electronically and that do not require significant degrees of experimenter interaction, such as Web pages. In some cases, the data are collected automatically, timed by the computer, with results sent back to the experimenter at the end of the assessment. In other cases, some form of remote audio and video connection is established so the experimenter can have a virtual presence and provide some degree of interaction with the user.

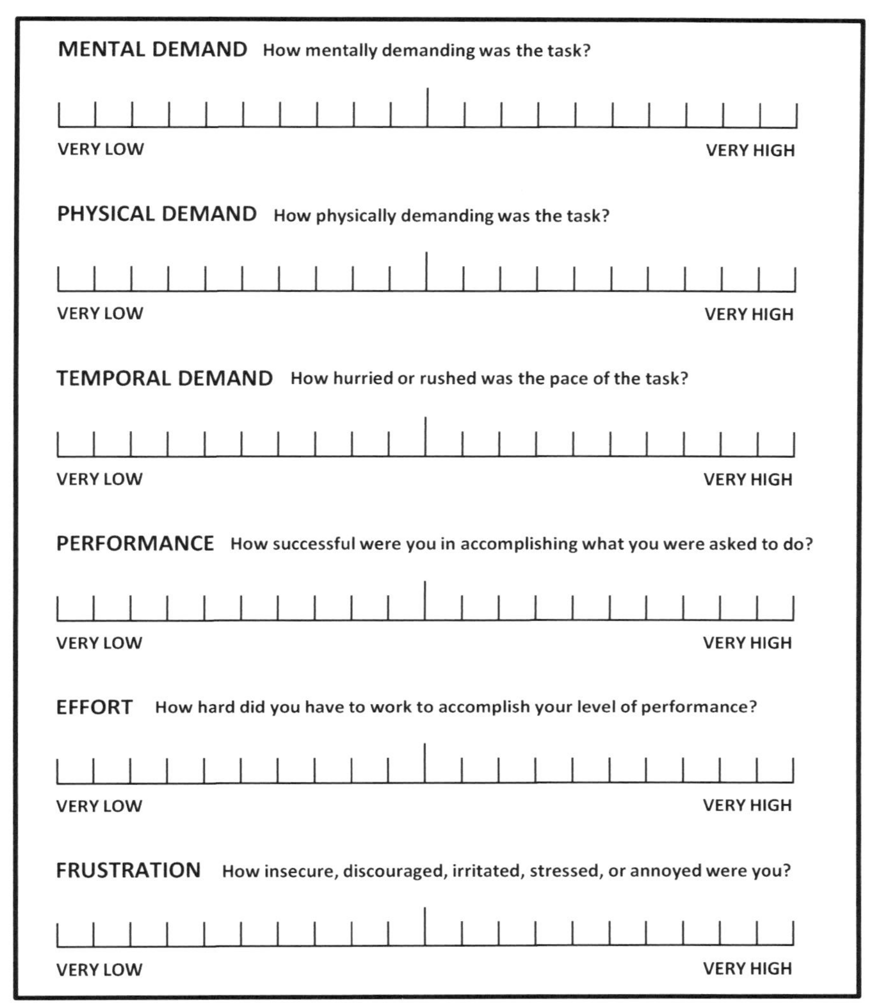

Figure 10: The NASA Task Load Index (NASA-TLX). Each of the 6 items is scored on a 100-point scale in units of 5. Overall score is computed by averaging the scores from the six dimensions. From Hart and Staveland (1988).

This kind of remote testing can be particularly useful if your users are distributed or if you have need of special populations that are not local, such as representatives in a specific call center whom you want to test. It can also have cost advantages, as users can be compensated less if they're not required to travel to your facility. Although this type of usability assessment can produce results that are on par with in-laboratory studies (Andreasen, Nielsen, Schrøder, & Stage, 2007) "You here, users there" remote testing can be more difficult because the practitioner does not have direct control over all the elements of the testing environment. Albert, Tullis, and Tedesco (2010) provided excellent additional guidance for this kind of testing.

4Bii) Remote Testing: You There, Users There

In other cases of remote testing, the practitioner travels to the user's site to perform the usability assessment. This might be done because the user population is specialized or there is a specific interface or piece of equipment that does not lend itself to being moved to the experimenter's location.

Sometimes the location itself is one of the variables of interest in the usability assessment. Such was the case in the assessment of the DSL self-installation kit. Some of the usability tests were performed in the users' homes to get a greater understanding of the kinds of usability issues that might arise in these kinds of naturalistic environments. The advantage of testing in the user's location is that you can see your product being used in a real environment and capture all of the relevant details. Again, these kinds of tests can be significantly more cost-effective if your users are being drawn from a specialized pool.

Logistically, this kind of testing can be more difficult. You don't have the "home-court advantage," so you must bring everything required to record data from your usability assessment. Further, there can be such a thing as *too much* realism if irrelevant distractors at the remote location significantly interfere with your user's ability to perform the task.

Take care not to confuse irrelevant distractors from relevant ones. If a user is constantly interrupted during performance of the task by the phone or coworkers, this may be a relevant issue to be aware of if, for example, your interface has significant memory loads or requires contiguous periods of concentration.

4Biii) Local Testing: You Here, Users Here

This is by far the most common kind of testing. Your users come to your facility and test the product in the environment that you have specified. In such a local setting, you have greater control over the test, and the setup is greatly simplified.

The amount of equipment that you will require to perform the usability assessment will vary significantly, depending on what you're trying to test. If your company is fortunate enough to have access to a fully equipped usability laboratory with one-way mirrors and audio/video recording equipment, you may want to take advantage of that. Even if your company does not have its own facility, these types of facilities are available to rent in most major cities.

Capturing and recording video and audio are some of the most prominent features of state-of-the-art usability laboratories. How important is it to capture audio and video? Video can be invaluable if you're trying to demonstrate specific problems to nonlocal team members through the use of highlight clips. However, remember that for each hour you test, you will have an hour of video that needs to be coded (which takes much *more* than an hour), and over the course of the full usability test, this can add up. If you run 40 participants, you have at least a week's worth of tape to watch and code. This can more than double the time necessary to review the data.

If you plan your data collection strategies carefully, collecting audio and video is not absolutely necessary. Over the course of the DSL self-installation usability assessment, we recorded thousands of hours of video on VHS tapes that we did not use extensively, because our real-time coding processes were more than adequate. Still, those data had to be protected (because of IRB requirements) and stored (not a significant problem in today's digital storage realm, but a much bigger problem when you're dealing with bulky physical media like VHS tapes). More important, it had to be managed, because for it to be valuable, we had to be able to find what we needed in this sea of tapes.

If you don't have access to a state-of-the-art testing facility, don't despair. The important thing to remember is that the testing environment only needs to be able to capture the relevant details necessary to understand the usability of your product or service. If you're testing a Web page, for example, you just need to be able to see what's happening on your user's screen. If you feel compelled to record the user's interactions, a number of extremely low-cost screen-recording software packages are available.

The kind of facility necessary for where the user will actually perform the tasks can vary greatly. If you're testing a Web site, it's likely that all your user needs is a computer (or tablet) and access to the Internet. Clearly, this kind of testing can take place in a conference room or an office, without any specialized equipment.

In the case of the DSL self-installation kit, a higher-fidelity testing environment was somewhat more critical. In the room, we needed the appropriate telephone jacks that were already provisioned with DSL signal. We needed to supply a computer that matched the operating system with which the user was familiar and a toolkit that contained the necessary tools for the installation that a consumer would be expected to have in their home. All of this needed to be provided in an environment that was a reasonable facsimile of a home location where DSL might be expected to be installed, so the installation was reasonably realistic.

Again, the important thing to remember is that the testing environment on the user side only needs to have enough fidelity to enable users to perform the necessary tasks in a sufficiently realistic fashion.

Do I need a fancy lab?

No! Although a well-equipped lab with one-way glass and video/audio recording equipment can make running participants easier and the collection of data more efficient, usability tests can be run in any number of environments with good results. If you've prepared adequately and can take good notes, expensive video and audio recording equipment is not required. So long as you have room for your user, the product or service being tested, and the experimenter to be present (in an out-of-the-way location), you have a more than adequate testing space.

In fact, it is often desirable to *no*t test in a fancy lab. This is true if the environment in which the interface will be used could have an impact on the user's ability to operate your product or service. For example, if I'm going to test a new interface that's going to be used on the floor of a sheet metal shop or foundry, testing in an environment that has similar noise and lighting conditions becomes very important. Although the interface might work wonderfully in a quiet, well-lit lab, it might suffer tremendously in an environment where there is significant noise, extreme lighting, and lots of other distractions.

4C) Define Your Timing Parameters

As noted earlier in the discussion of efficiency measures, time is one of the most commonly used metrics of efficiency. Although the measurement of time on task seems relatively straightforward, it is very important to carefully define both start and stop times in a way that can be translated into operational practice. If multiple members of the team are taking time measurements and are not all using exactly the same criteria, your efficiency data will be impossible to interpret.

Use specific actions to define the start and end times. For example, if you're using a computer to run the experiment, start timing when a user presses the first button, or keep the screen blank and start timing when you unblank the screen. The key is to explicitly define your timing landmarks and then *be consistent.* Good ending landmarks include when the user presses a "finished task" button, or says he or she has completed the task (whether or not the user actually did finish), or when the user writes down an answer, if that is part of the task requirement.

Some practitioners suggest that the experimenter can make a judgment about whether a user is making progress and is likely to complete a task. If the user is not likely to finish, the experimenter "calls it" (Rubin & Chisnell, 2008). However, this isn't advisable if you don't have extensive experience, because the criteria for "calling it" can be too subjective. Use the data-driven upper time limit that you established in section 4Ai, and let the user work until that point.

During the DSL self-installation, we started timing when users touched the self-installation box and ended timing when they launched the browser with an Internet connection, or when they self-reported that they were finished or couldn't finish.

4D) Define Your Testing Material and Equipment

In addition to your facilities, you'll need a number of other elements in order to perform your test. It's important that you define all of them before you begin to ensure that you'll have everything you need.

4Di) Product/Service Being Tested

If you're testing a physical product, you will, of course, need at least one of those. It's often important to have several of the products on hand in case one becomes broken or disabled during the test. If you are testing a Web page, you'll need some way for the users to access it, such as a computer or tablet, depending on the goals of your test. Make sure to match the computer's operating system and browser with the preference specified by your user, or you may not be testing what you think you are.

In the case of DSL, we needed a self-installation kit that was identical to the kit received by real users. To obtain valid results, it's important that each user be presented with *exactly* the same equipment and setup. During the early rounds of DSL self-installation testing, we simply obtained a fresh, unused self-installation kit for each participant we anticipated testing. As it became evident that we would be testing hundreds of users, this option became cost prohibitive, so we procured enough material that we could repackage the entire kit, including the outer box, in a way that made each user believe that he or she had received a fresh, new kit. Because the installation process installed a number of pieces of software and drivers on the computer, it also meant that at the end of each test, we had to completely reformat the computer's hard drive and reinstall a hard drive image that put the computer back into a known state.

Even for tests that are simply examining the usability of Web pages, the browser cache should be cleared, the history wiped, and the browser restarted so new users don't get the benefit of actions taken by previous users.

4Dii) Support Material and Services

Make sure that you have the materials to support all of the test scenarios in your usability assessment. This includes things like usernames and passwords, fake e-mail addresses (so users can protect their privacy by not having to use their personal e-mail address), account numbers, required tools, training material, and any other items that are necessary for your users to complete the task. Use the output from your cognitive walk-through to help identify all of the required elements. It also may be necessary to prepare "fake" support services if you'll be allowing users to take advantage of such services but won't be utilizing currently operating help facilities.

In testing the DSL kit, we needed to create not only the usernames and passwords the participants would need but also the welcome letters that contained that information, because finding those details was an important part of the task. More important, we had to make sure that these credentials were loaded onto the real DSL password servers so a user could actually log on and use the system. Because the users' e-mails would be collected during the registration process, we also needed to generate fake e-mails for the participants to use, along with alias identities (e.g., John Smith at 123 Main Street), so they didn't have to enter their own information into the system. Because we were able to use real backend systems, we didn't have to create fake help desk services. If a participant was having trouble, he or she could call the listed support number using the fake credentials we had supplied and talk to real support personnel (which also provided some extremely valuable data on how to best help users with common usability problems).

4Diii) Test Materials

Identify all the test materials you'll need during the test so they're on hand and so those reading your test report will know what is being measured and how. These items can include surveys (e.g., System Usability Scale, After Scenario Questionnaire, NASA-TLX), demographic questionnaires, consent forms, and debrief forms. These should be completely constructed. It's not sufficient to simply say, "Will have a demographic questionnaire" – you should have all the questions that you intend to pose to the users identified as part of the test plan.

You'll also need to include whatever is necessary to compensate the users (gift cards, checks, cash) and any accounting mechanisms necessary to support this compensation, such as participant acknowledgment-of-payment forms. Don't forget to specify items that will be there for the participant to use during the test. This includes things like pens and pencils, notepads and clipboards, and refreshments (if any are to be provided).

In addition, specify the basic data collection materials that you'll be using, such as notepads, laptop computers, stopwatches, automated or semiautomated data collection tools (such as Morae or Noldus Observer), and any spreadsheets that you'll be using to capture users' behaviors. The idea is to create a list that is sufficiently detailed that it can serve as a checklist. If all the items on the checklist are available, the test can be run.

4Div) Procedures

With the users' tasks specified during preparation for your cognitive walk-through (see section 3E on task definition), now it's time to specify exactly how the test will be conducted. First, determine how long the total test session will take. You tend to get the best user performance if you keep the entire test session under 90 minutes – longer than that, and it can become difficult to maintain the users' attention. Of course, some situations demand significantly longer test times, so a 90-minute limit is not a hard and fast rule.

If the number of tasks that you have defined during your cognitive walk-through cannot be done within this timeframe, consider multiple sessions or trimming the task list to a subset that fits within the requisite time. If you choose to trim your list, do so with great caution, as the tasks you identified for the cognitive walk-through have already been defined as important and are likely worthy of user testing.

Make sure to leave enough time between participants so you can save your data, reset the equipment, and debrief with your team regarding any unexpected events.

For the DSL testing, we had an upper limit of 90 minutes and included an hour to restore the laboratory and computer to its original state.

Next, define the procedures for tests that go exactly as planned. This includes greeting the participants and checking them in to your facility, getting them situated in the laboratory, obtaining their informed consent, and giving them the end-of-test debrief.

Procedures regarding the delivery of tasks should also be described. Will these tasks be given to the participants orally, or will the participant have a set of written instructions? Communication protocols for the user should also be specified. It is highly recommended that you have a minimum of interaction with the participant. It's good practice to tell the user before the test starts that you will be communicating with them only between tasks, or to tell them to move to the next task. Set the expectation that you won't be able to communicate with them at any other time or provide assistance. Avoid the temptation to provide hints to users during the testing to help them get through a task. Doing so will only demonstrate the quality of the hints you give, and any information about the usability of your product or service will be lost.

Scripts should be used to ensure that you're providing every user with the same information, especially if there will be more than one experimenter running test participants. If your users end up asking for help anyway (as many do), simply say, "Do whatever you think you should do to <insert the name of the task they are doing here>."

The procedures should also identify any contingency plans that are in place in case the test does not go as planned. For example, if there is an equipment failure, how long will you wait before you release the participant? It is also important to specify when you will intervene in any given task. Generally, it's best to let users perform the task as they see fit within the time constraints that you have specified earlier in section 4Ai. Intervention should be undertaken only if the user or the equipment is in danger.

These are the parameters we specified for the DSL self-installation, and it was fortunate that we had. In one case, we intervened with a laboratory power shutdown when the user managed to open the computer chassis without unplugging it and was preparing to put a screwdriver on the motherboard to "pry off the PCI slot cover" (her words).

Consider what information you'll want to collect from your participant after the test is finished. Sometimes this can be performed in an unstructured manner simply by asking the user, "So, how did it go?" In this case, the participant will self-identify those items that he or she felt were most important during performance of the tasks.

The difficulty with this approach is that often, users may have forgotten key points that they wanted to convey, or they are simply ready to finish and leave and so they don't provide any significant input. It's better to have a series of questions that strike at the heart of the difficulties users may have encountered. This means that unlike the rest of the very structured test plan, your debrief questions may need to be constructed on the fly during the testing.

For example, during the DSL testing, we observed that many users spent a significant amount of time working with the DSL filtering devices that plug into the wall jack. Eventually most users got it right, but only after significant effort. During the debrief we used this opportunity to ask users what it was about the installation of the filter that seemed to make it difficult. In this way, we were able to focus users on a product inefficiency that we observed and helped remind them of the task in which they struggled.

Often, people who have a vested stake in the product want to observe the usability testing. This can include product developers, product managers, programmers, and other usability professionals. Having observers in your usability tests is generally a good thing because they can get first-hand knowledge of the kinds of usability issues that their potential customers will face. That said, you must develop procedures for managing these observers. Determining where they will sit, how you interact with them during the test, and whether you will allow them to provide guidance during a usability test are all important to consider. Make sure that your observers arrive well before your first participants. Nothing makes a user more nervous (and contaminates your results) than seeing a large group of people file into the observation room right before the test.

During DSL testing, we encouraged members of the product development and management teams to attend the usability tests, but we set expectations by letting them know beforehand that they were there strictly to observe and that, although we were happy to discuss anything with them at the end of the day, the test procedures were too demanding to allow for interactions with them during the actual testing. Table 8 shows a list of the major elements that you'll need to define as part of your test plan.

If your procedures are well written, a practitioner who was not involved in the planning of the test should be able to step in, follow your protocol, and run the entire usability test without having to make any operational decisions about how to perform the usability assessment. Keep this goal in mind as you consider the details necessary to construct a good test plan.

4E) Verify Your Users

When you were preparing for your heuristic evaluation, you defined the population that would be using your product. Now that you've finished your heuristic assessment and cognitive walk-through, it's time to make sure that that user definition is still accurate. Check with your marketing colleagues to ensure that the target population hasn't changed and that your preliminary assessments didn't identify any new classes of users that should be considered. The definition needs to be exactly right because the next step is to recruit actual users.

4F) Obtain Approval (IRB)

In response to the unethical treatment of human participants in a number of studies in the 1950s through the 1970s (notably the Tuskegee study, Corbie-Smith, 1999; and the Stanford Prison Experiment study, Zimbardo, 1973), the National Commission for the Protection of Human Subjects of Biomedical and Behavioral Research issued the Belmont Report (1979). This report detailed the requirements for the ethical treatment of human participants in research studies. Later codified into law by the U.S. Department of Health and Human Services (45 CFR 46), these rules, often referred to as the "Common Rule," mandated that human participants and their data be sufficiently protected when participating in biomedical or behavioral research studies.

Table 8: A List of the Major Elements That You Might Need to Define as Part of Your Test Plan

Product/Service Being Tested

- The actual product or service of interest, and a spare (if available)
- The computer and operating system if the user requires access to a computer or the Internet

Support Material and Services

- Usernames and passwords for the systems you will be accessing
- Fake e-mail addresses, if required, so the users don't have to use their own
- Account numbers necessary to perform the tasks
- Tools necessary (screwdrivers, pliers, etc.) for the user to perform the task
- Supplemental software that might be required (e.g., a word processor, an e-mail client)
- Training material
- Help manuals
- "Fake" support services, such as help desks

Test Materials

- System Usability Scale survey
- After Scenario Questionnaire
- NASA-TLX
- Demographic questionnaires
- Consent forms
- Debrief forms
- Office supplies (pens, pencils, notepads, recording media)
- Stopwatches
- Data collection computers
- Data collection spreadsheets
- Automated or semiautomated data collection tools (e.g., Morae or Noldus Observer)

Procedures

- Establishing the timing of the test, including total test time and time between tests
- Getting the user checked in through security and situated in the lab
- Obtaining the user's informed consent
- Describing the course of the test and its goals to the user
- Determining how the tasks will be described and delivered to the user
- Determining how and when you communicate with the user
- Creating scripts for all communications and instructions
- Defining contingency plans for equipment failures
- Defining contingency plans for user no-shows
- Setting intervention strategies for situations that are dangerous to the user or the equipment
- Establishing post-test information collection procedures (debrief)
- Determining how you will manage test observers
- Creating team debrief strategies for use at the end of the test

Note: *All* of the procedures will be required and are presented in the approximate order of their execution.

The implementation of these rules ensures that participants understand the risks and benefits of the study in which they are participating, and that no harm comes to them during the course of their participation. All research plans must be evaluated by an Institutional Review Board (IRB). Any research that is funded by the U.S. government is required to comply with the Common Rule and must be evaluated by an approved IRB.

Nearly all academic institutions receive federal funding, and so the IRB process is well integrated into research programs in academia. To maintain conformity and minimize confusion, most universities require that all research involving human subjects comply with the Common Rule, even if it is not funded by the federal government (Gunsalas et al., 2007). If you are conducting usability assessments for the U.S. government or at a university, you will need to seek approval from your institution's IRB before you can conduct your usability assessment study.

The situation in the industrial sector is not as clear-cut. The Common Rule contains exemptions for certain kinds of research that deal with anonymous surveys, educational research, or studies that use existing data (45 CFR 46 101b), and notes that studies involving little or no risk to the participants are exempt from IRB review. Universities have determined that the IRB makes the determination of whether a study involves minimal risk and is therefore exempt, but such is not the case in many business settings. Often, large corporations or those that receive federal funding function much like their university peers and require IRB approval for all human subject testing. Smaller companies may feel that usability research is of minimal risk and does not require IRB approval.

If you are testing children or other vulnerable populations, such as cognitively impaired individuals, obtaining approval from an IRB becomes substantially more important. Consult with your company's legal department to determine if formal IRB processes need to be adhered to for the types of studies you're conducting. If you intend to publish your work, many of the top journals require that participants fully consent and are ethically treated, and require IRB documentation to that effect.

Regardless of whether you obtain formal IRB approval, always provide your participants with an informed consent that describes (a) the nature of the study, (b) the potential risks to the participant, (c) the potential rewards (e.g., compensation), and (d) the right to end participation in the study for any reason at any time. It should also describe what steps you are taking to ensure that the privacy of that person's data is maintained, which can be particularly important if video and audio recordings are being made (Waters, Carswell, Stephens, & Selwitz, 2001).

4G) Suggested Reading

- *Usability Inspection Methods*, by Robert Mack and Jacob Nielsen. John Wiley & Sons, 1994.
- *User-Centered Design: An Integrated Approach* by Karl Vredenburg, Scott Isensee, and Carol Righi. Prentice Hall, 2002.
- *Beyond the Usability Lab: Conducting Large-Scale Online User Experience Studies* by William Albert, Tom Tullis, and Donna Tedesco. Morgan Kaufmann, 2009.
- "Research ethics meets usability testing" by Susan Waters, Melody Carswell, Eric Stephens, and Ada Sue Selwitz. *Ergonomics in Design, 9,* 14–20. 2001.

CHAPTER 5: PERFORM THE USABILITY TEST

Now that you have performed your heuristic assessment to gain an understanding of how your product works with the population of users you've identified, defined a set of tasks to use in your cognitive walk-though, and created a full and complete test plan that details every experimental detail, you're ready to begin the assessment of your product or service with real users. Now is when you'll put your test plan to work, following it to the letter in the setup and execution of your usability assessment.

5A) Recruit Your Users

Of course, you'll want to recruit users who have the right level of knowledge, expertise, and experience. But where do you find these users? As noted in section 3C, you shouldn't recruit from inside your company, as those users will likely have specialized knowledge that might aid them in the performance of certain tasks with your product. Don't recruit friends and family, even if they meet the demographic profile. They typically don't want to hurt your feelings and they may be biased in how they report any difficulties they might encounter in the use of the product or service.

There are many ways to recruit the participants you need for your usability assessment. One of the easiest ways, albeit the most expensive, is to use a marketing research or recruiting firm. You supply the firm with the exact demographic characteristics required for your test, and, through their recruiting networks, they supply a matching set of users. Costs can range from $100 to $500 per participant, depending on the specificity of your user profile. This is the method that we utilized extensively for recruiting DSL self-install users, as it allowed us to focus on the testing preparation rather than recruitment.

A significantly less expensive way to recruit participants is to use Craigslist or newspaper ads. The cost for placing the initial add is fairly reasonable, but there can be a significant time investment in screening those who have answered your advertisement. In a recent set of studies, we placed a set of ads that had specific user characteristics listed as requirements. However, at the first in-person meeting, nearly all of the potential participants had to be disqualified because they hadn't fully represented themselves when they had first responded. Often, repeated advertisements are necessary, as the number of participants who respond to a single ad can be low, and those who do respond are frequently unqualified for your studies.

If your product lends itself to remote testing (e.g., a Web site), services such as Amazon.com's Mechanical Turk can be excellent resources for recruiting users. The cost for the services can be exceptionally low (as low as $.25 per user), but demographic matching can be difficult and the commitment of your users questionable (Paolacci, Chandler, & Ipeirotis, 2010).

If you're searching for users from specific, hard-to-find demographic categories, seek out organizations where potential users with those demographics tend to congregate. For example, professional societies can be a good source of users with a specific skill set. Nursing homes can provide older users and people with cognitive and physical limitations. State-run schools for the blind and visually impaired and schools for the deaf and hearing impaired are excellent places to recruit users who have specific disabilities. More general populations can be found at civic clubs, such as the Rotary club or the League of Women Voters. Be cautious when selecting users from these kinds of groups, however, as they tend to be highly motivated and can skew your results.

One excellent place to recruit users is your own customer base. Having already demonstrated that they use your products and services, your customers are likely to use newer products and services made by your company. In early tests of the DSL self-installation kit, we recruited users who were subscribed to our dial-up service because they were likely customers of the new high-speed Internet service offerings.

Some firms build and maintain their own direct recruiting lists. Taking a similar approach can provide you with a ready pool of users of a known demographic, but construction and maintenance of the list is a time-consuming endeavor. More important, it can lead to the cultivation of so-called professional users who are tested again and again with your products and services so that, over time, they begin to look more like employees than real users, and so using them is less than ideal.

If your product is for general use and the target population is everyone, make sure that your recruited population is representative of the range of user characteristics of the general population, including age, gender, and ethnicity. In some areas of the country this can be easy to do, but in other, more homogeneous areas, the task may be more difficult. If you're in an area where the population is too homogeneous, consider testing in a part of the country where broad demographics can be easily recruited.

Testing with participants who are not fully representative of the diversity of your actual user population makes it difficult to generalize the results of your testing; you end up having confidence in the usability of your product for that specific group but not for the population you really wanted to assess. Check your recruited sample for the required diversity *before* you test, as it's always less expensive to recruit the first sample correctly than to resample when your first group of users is found to be insufficiently diverse.

There are times when you won't be able to recruit the exact demographic group that you need. Sometimes it's simply a matter of cost. For example, you might be working on a piece of equipment designed for use by nurses in rural Africa and don't have the budget to travel there to perform usability tests. Or perhaps there is a very limited number of the exact demographic you need for your product, and their availability is greatly limited. If you're designing a product to be used on the space station, you might not be able to get ready access to the very limited number of astronauts to test your product.

In these cases it is common to recruit users who may not be the exact group who will eventually use your product but who share enough critical characteristics of your final end-users that the use of so-called proxy users is an acceptable alternative. That said, whenever possible, test with your exact user population.

5Ai) How Many Users Do You Need to Recruit?

The number of users needed for a simple usability test should be a straightforward matter. Unfortunately, it is a contentious issue even within the usability community. Some usability practitioners (Nielsen, 1993, 2000; Virzi, 1992) believe that you can uncover the majority of usability issues in a product by testing only 5 users. On the opposite end of the spectrum, there are those who believe that significantly more users (> 20) are necessary to identify the majority of usability issues (Faulkner, 2003; Spool & Schroeder, 2001). The National Institute of Standards and Technology Voter Performance Protocol recommends at least 100 users in the assessment of voting systems (NIST, 2007). A host of other sources suggest numbers between these two extremes (Hwang & Salvendy, 2010; Lewis, 1994; Nielsen & Landauer, 1993; Woolrych & Cockton, 2001).

Does geography matter?

Geography can be very important in the marketing and sale of a product. You don't market space heaters in Death Valley, California, or air conditioners in Nome, Alaska. People have different preferences, so marketing and advertising executives go to great lengths to employ data-driven strategies to ensure that they're marketing the right product to the right group.

It's tempting to think that usability is similar. It's not. Despite evidence that usability can vary across countries (Clemmensen et al., 2007), data from my laboratory has shown that there are no significant variations in how different users across the United States view the usability of a certain products or services. Given the vast variations that marketing professionals see, how could there be such a difference with regard to usability? The reason is simple; marketing examines preferences, whereas usability examines skills. So long as the demographic you are testing is matched appropriately, users in different geographic locations will not show large variations in how well they can use a product. What this means is that if your user profile says that your product will be used by 30-40-year-old engineering professionals, where those engineering professionals are from is essentially irrelevant.

Some of the original work in understanding the number of users required for usability testing was performed by Virzi (1992). He described a model whereby the number of unique usability problems found could be described by the formula $N(1-(1-L)^n)$, where N is the total number of usability problems in a product, L is the likelihood of a specified usability problem being found, and n is the number of participants. N can be set arbitrarily (100 is convenient), as the result can then simply be read as a percentage of problems found. Based on data from a number of usability tests, Virzi determined that the *average* likelihood that a user would find a specified usability problem (L) was approximately 0.3. This results in a curve shown in Figure 11, where approximately 85% of the problems are found by testing with 5 users.

As can be seen, however, the formula is sensitive to the value of L, with other researchers suggesting that the optimal value is closer to 0.1 (Hertzum & Jacobsen, 2001; Law & Hvannberg, 2004). This would result in needing almost 20 users before finding 85% of the usability problems in a product. Further, Faulker (2003) demonstrated that variability in the selected user sample and likelihood that a problem will be found can cause the percentage of problems found with 5 users to range as low as 55%.

Schmettow (2012) suggested that the Virzi model significantly underestimates the number of participants required, and the curve based on Schmettow's LNB_{zt} model, shown in Figure 11 for comparison, suggests that at least 56 users are required to reach the 85% threshold. Although the academic literature may seem hopelessly conflicted on the number of users that are required, the differences are based in large part on assumptions regarding the kind of test being run, the true likelihood of discovering a problem, the complexity of the experimental design, and the product being tested.

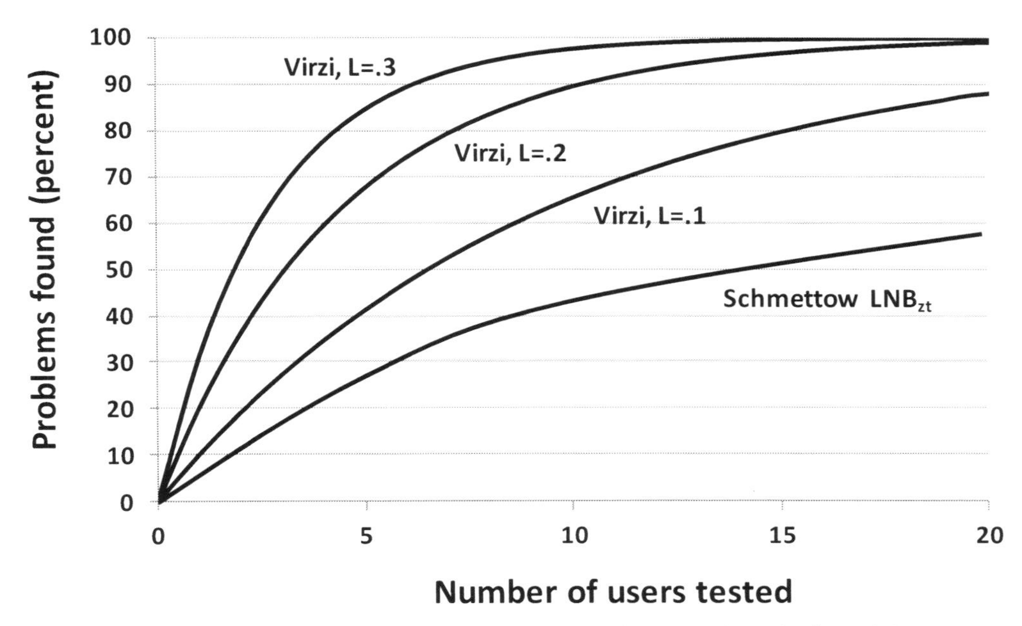

Figure 11. The proportion of problems found using Virzi's original model (1992) for three values of L (the likelihood of a problem being found), based on the number of users tested. A model proposed by Schmettow (2012), which requires significantly more users, is shown for comparison.

Participants are one of the major expenses associated with usability assessment, so minimizing the required number of users is important. There is general agreement that formative tests require fewer users than summative tests, and that between 5 and 10 users are sufficient for a formative test, especially if you are planning on testing in an iterative fashion. If your product has significantly different user groups, you may need to test 5 users from *each* group to obtain results that are meaningful for your entire user population.

Because you are trying to achieve greater statistical reliability in the measurement of usability metrics in a summative test, more users are generally required in this case. Generally, 20 to 30 users is a typical sample size for summative tests, although this number can be much higher if there is high variability in the sample. Macefield (2009) provided a helpful practitioners' guide for sample size selection in usability assessments, and the work by Sauro and Lewis (2012) contains excellent coverage of sample size calculation for a wide variety of testing circumstances. Both sources are listed at the end of this chapter in Suggested Reading.

Over the course of the DSL self-installation kit development effort, we tested more than 1,000 users. Those users participated in a large number of formative studies, usually between 5 and 10 users, and a series of larger summative studies, in which we tested between 20 and 30 users, depending on the exact test conditions.

5B) Conduct the Pilot Test

The pilot test is your opportunity to shake out any potential problems with the test plan. The pilot test ensures that all of your tasks work as expected, that the timing procedures are appropriate, that your scripts work as planned, and that all the operational details of the test have been identified and work in concert as part of the complete usability assessment.

Except for the number of participants, the pilot test looks just like the full usability test in its execution. It's a full dress rehearsal, so you can't leave anything out. The big difference is that the pilot study should be conducted with only a few participants. Make sure to test at least two users during the pilot to ensure that you have allotted sufficient time between participants to reset the equipment and make preparations before the next pilot participant arrives. Use the same user demographic as you'll be using for your actual test, or the problems you discover during the pilot testing may not be fully representative of what you might see with the appropriate demographic.

The pilot test is also your opportunity to hone your moderator skills. As part of the test plan, you'll need to have developed scripts for all of the important interactions (greetings, consent, task assignment, debrief). However, as with any social interaction, there will be unscripted moments. At those times, treat your participants with kindness and professional consideration, remembering that you are testing the usability of the *product,* not the skills of the *user.* When users fail to complete a task, they are often embarrassed, and it is important to put them at ease by reminding them that you are interested in finding deficiencies in the design of the product and that their assistance in this matter has been invaluable.

Be aware of how your body language and facial expressions may convey information to the user about their performance. During the DSL testing, we told participants during the initial briefing procedure that we were testing how good the design of the self-installation kit was and that any difficulties they encountered were due to deficiencies in that design and didn't reflect on them (the user) in any way.

One of the biggest mistakes I have observed during pilot testing is that the test isn't run in a sufficiently rigorous fashion, and that significant shortcuts are taken. Certain steps are dismissed with a wave of the hand and the suggestion that "we'll take care of that during the real test." Run the pilot *exactly* as you plan to run the real test – only in this way will you uncover any weaknesses in your test plan. Make sure that any audio and video recordings you made were actually recorded, and don't forget to reset all of your equipment and clear the data spreadsheets after the pilot study is complete so you don't mix pilot data and data from the full test.

Resist the temptation to use the pilot study data "if everything goes well." Assume that you will be making minor adjustments to the protocol and test plan, and treat the pilot test as the last step in the planning exercise before the actual usability assessment. Finally, don't assume that your pilot test will catch everything. You may still need to make minor adjustments during the actual testing as the need arises. That said, if you've been diligent in the construction of your test plan, and rigorous in execution of your pilot testing, further modifications should be minimal or unnecessary.

5C) Conduct the Full Test

The pilot test is complete. The test plan has been modified based on findings from your pilot test. You are now ready to conduct the full usability assessment. The night before testing begins, perform a physical walk-through of the lab. Not unlike the commercial airline pilot who walks around the aircraft before each flight, you want to make sure that your usability assessment is also ready to "fly." Use the test plan as your guide. Is all your paperwork ready? Is all of the equipment operational and reset from the pilot testing? Is your compensation ready? This final walk-through will help you mentally prepare for the next day's events.

When the test participant arrives, execute the test exactly the way you did in the pilot study. Following the debriefing and dismissal of your participant, secure your data. After each participant, make a backup copy and ensure that your recordings are also backed up, if you've made any. There's nothing worse (or more costly) than running several participants and then losing the data due to equipment or computer failure. Once your data are secure, begin your reset procedures, getting the equipment back to a state so it's ready for your next user.

If you set aside sufficient time between the tests (as you should have), debrief with your fellow practitioners to discuss any testing anomalies and unexpected or particularly significant results. At the end of the day, a short roundtable discussion with your observers should be undertaken so you can summarize the day's results and ensure that your observers understand what they've seen and have been afforded an opportunity to provide input to you.

Just because you've run a rigorous usability assessment doesn't mean that you've captured all of the problems associated with your product. Because of sampling issues, it is clear that no single usability assessment is likely to capture *all* of the usability problems from which a product suffers. Does this mean that the results from usability tests are not valuable? Certainly not. What it really means is that results from a single usability test must be viewed with some degree of caution. It also suggests that using robust usability assessment methods is extremely important and that employing an iterative usability assessment methodology will also increase the effectiveness of the assessment, given that repeated testing is likely to uncover a slightly different set of problems than those encountered in a single test alone. By testing more than 1,000 users for the DSL self-installation kit, we had high confidence that we had identified the significant problems associated with the product.

5Ci) Variations on a Theme: Think-Aloud and Codiscovery Methods

One common variation to the usability assessment methods described earlier is the application of so-called think-aloud protocols in an effort to have users themselves provide an ongoing assessment of what they are thinking while they are trying to use the product. Its origins can be found in Ericsson and Simon's (1980) explorations of using verbal utterances as data that can be analyzed. Modern application of the think-aloud protocol in usability assessment does not rigorously follow this psychological model (Boren & Ramey, 2000), but many researchers extol the benefits of using this kind of protocol (Jaspers, Steen, van Den Bos, & Geenen, 2004; Nielsen, 1992b, Nielsen, Clemmensen, & Yssing, 2002; Van den Haak, de Jong, & Schellens, 2004).

The basic idea is fairly simple. As users attempt each task, they are encouraged to provide a running commentary of what they're thinking, what problems they're encountering, and how they're solving these problems. In this way, identification of the problem and its underlying cause is not left to the subjective assessment of the practitioner who is observing the user do the task; rather, it is explicitly stated by the users themselves.

At the beginning of the test, the experimenter tells a user that to describe what he or she is thinking while performing the task. These verbalizations are not meant to be a description of the physical actions that are taking place but, rather, should be descriptions of what the user is *thinking* while doing a specific task. For example, a user statement of "now I'm picking up the DSL modem to plug it in" isn't particularly useful, as this fact can be readily observed by the experimenter. However, statements such as "I'm looking at the back of the modem and I'm confused about what needs to get plugged in the back here, because all of the ports are unlabeled" would provide significant information to the experimenter. Nielsen (1994b) reported that the think-aloud protocol is as efficient as other usability assessment methods, so no additional participants are required when it is used.

Are the results reproducible?

When you're working with an engineering team, they like to see reproducible results. An electrical engineer will tell you that if she puts 2 volts in *here*, she can reliably measure the 2-volt output *there*. Unfortunately, as a behavioral science, usability assessment does not exhibit the kind of reliability in results that we typically see in systems governed only by the laws of physics. Usability assessment looks much more like the detection of computer program bugs. We know there are bugs in the computer program. The most egregious ones can be detected quite reliably, whereas some of the more subtle bugs can be detected only under a very exact set of circumstances. If usability assessment methods are applied consistently and rigorously, it is likely that the most egregious usability deficiencies will be detected. However, the reliability of usability error detection declines as the severity of the error decreases.

A cautionary tale regarding the reliability of usability assessment results can be found in the work of Mölich and his colleagues (1999, 2004, 2008). In these studies, a number of independent usability laboratories performed usability assessments of the same product. The authors found that there was extremely low agreement on all aspects of the usability assessment across the labs. The labs employed different methodologies, used vastly different numbers of users to make evaluations, and selected widely varying tasks to make the assessment. Of greatest concern, however, was the fact that there was little overlap in the problems found by each of the laboratories. For example, in one of the studies (Mölich, Ede, Kaasgaard, & Karyukin, 2004), nine teams evaluated the usability of the Hotmail program. In those nine evaluations, 75% of the problems reported across all tasks were found by only a single laboratory, and not a single problem was reported by all nine teams.

One of the major difficulties with think-aloud protocols is that users tend to stop talking, particularly when the task becomes difficult. It is incumbent upon the experimenter to gently remind the user to "keep speaking about what you are thinking" if the user's verbalizations begin to wane. Erickson and Simon (1980) recommended that users should speak as if they were in a normal conversation, which is about 100 words per minute. These verbal utterances, rather than the practitioner's assessment of what the problems are based on the user's actions alone, become the basis for the identification of problems.

A technique called *codiscovery* can help to solve the difficulties of keeping users speaking via the think-aloud protocol. In a codiscovery methodology, two users interact with an interface and describe *to each other* the difficulties they are encountering. Because another user is present, the task is significantly more natural and appears to be normal conversation between two people. Although codiscovery can take place independently of the think-aloud protocol, the two techniques are often described in the literature as occurring together (Adebesin, De Villiers, & Ssemugabi, 2009; Kemp & Van Gelderen, 1996; Siau, 2003).

Even if you aren't formally employing a think-aloud protocol, some users (particularly those who have participated in other usability studies) may spontaneously describe what they're doing and the difficulties they are encountering as they perform a task. In these cases, it's best to let the users continue to speak, as they have already demonstrated that this is what makes them comfortable.

The think-aloud protocol is quite popular, but it should be used with an awareness of the impact it has on your testing and results. The process of thinking aloud can often aid users in solving problems that they might not otherwise be able to solve if they weren't employing this technique (Berardi-Coletta, Buyer, Dominowski, & Rellinger, 1995). Therefore, it might *positively* impact users' performance, boosting success rates in ways that might not be seen in the field. On the other side of the coin, talking while you're performing a task can add significant additional cognitive load (Preece et al., 1994), and the effort associated with talking and performing the task simultaneously might *decrease* the user's efficiency.

Because of these potential difficulties, we did not employ formal think-aloud or codiscovery protocols when testing the DSL self-install kit. The richness of the data from user behaviors alone appeared to be sufficient. The exception to this general practice was when we went to people's homes to perform usability evaluations. In that setting, it seemed to be more socially acceptable for the users to be able to talk to us as they performed the installation. That said, it has been demonstrated that think-aloud protocols provide data that are comparable to protocols in which it is not used, and that the added richness of the data can be beneficial (Virzi, Sorce, & Herbert, 1993).

Think-aloud protocols seem to have some problems. When *should* I use them?

Think-aloud protocols are perfect when you are more interested in the user's thought processes and mental models than you are in their actual performance on a set of tasks. By gathering users' spoken thoughts as they proceed through a task, you can gain a greater understanding of how they think the interface *should* be working and where and why they are having difficulties.

In a recent study, we were evaluating the usability of a novel voting method that allows a person to vote for the same candidate more than once if he or she so chooses. You can imagine the difficulties with using an interface like this, which is so different from the standard model of voting that most people have. To get a sense of the kinds of difficulties users might be experiencing when voting with this interface, we first employed a think-aloud protocol to gather data about what the users were thinking while performing the task. With the data from the think-aloud protocol, we could make minor modifications to the interface design and then proceed with a full-scale standard usability assessment.

Think-aloud protocols can be very valuable, but because of the interference they can cause during those critical times when users are performing their tasks, think-aloud should be used with great caution and in limited circumstances.

5D) Report Your Results

Once your usability assessment is complete, you'll need to report the results. The purpose of the report is not only to document what happened but also to convey information to the team about the product's overall usability and identification of the usability problems encountered by the users. If you're reporting the results of a summative usability assessment, the majority of the results will be the quantitative presentation of effectiveness, efficiency, and satisfaction measures. If you're reporting the results of a formative usability assessment, you'll likely report these three quantitative measures, along with some additional qualitative assessments of the usability problems that need to be fixed and how best to do that.

A summative usability report should contain a number of standard items, as defined by the Common Industry Format (CIF), and documented in ANSI/INCITS 354-2001 and ISO/IEC 25062:2006. These two standards aim to promote a common format to report usability data to facilitate easy comparisons across multiple tests and/or usability test groups. In essence, they define the requirements for a *usable* usability report. The CIF does *not* specify what tests you should run, how to run those tests, or how to interpret any of your data. It does assume that you have collected the quantitative usability measures of effectiveness, efficiency, and satisfaction as defined by ISO 9241-11. Also note that the CIF does not include a section for recommendations.

How do I get my results implemented?

If you want to get your results implemented, it's important that the team have full trust in the output of the testing that's presented in your report. Be rigorous in your data collection and analysis, and don't engage in speculation. Let the data do the talking. Be diplomatic in your discussion of usability deficiencies, as the product that you tested was the work of others, and they are likely to be the ones who will have to fix it.

Being diplomatic, however, does not mean that you can't provide honest assessments of the performance of the product from a usability standpoint. In many large companies, even though the value of usability has been amply demonstrated, it's still necessary to be a team player. In smaller companies, you may need to ease into the process. Remember, you have to be able to convince people that your data are sound and that the remedies that you are proposing will improve the usability of the device in the next version. Successful usability practitioners are not only technically competent, they're accomplished in their people skills (Kortum & Motowidlo, 2006).

Several customized versions of the CIF, most notably for electronic health record systems (Schumacher & Lowry, 2010) and voting systems (NIST, n.d.), have also been developed by the National Institute for Standards and Technology (NIST). Required elements in the standard CIF report are shown in Table 9. NIST provides a sample of the completed CIF report on its Web site at http://zing.ncsl.nist.gov/iusr/documents/diarymate_v32.htm.

The CIF was developed specifically for summative usability assessment reports. There is no standard at this time for formative usability reports, although one is being developed for inclusion in ISO/IEC 25066. Given that a formal reporting format does not exist for formative usability assessment reports, many practitioners use the framework specified in the current CIF and modify it to include the kinds of data collected during formative usability assessments (e.g., Masip, Oliva, & Granollers, 2014). Most notably, this results in the inclusion of a section (in Results) that details the frequency and severity of specific kinds of errors and provides more detailed descriptions of those errors. Most formative reports also include a recommendation section to help guide the designers in rectifying and mitigating the most severe errors noted in the results section.

5E) Postlaunch and Postmortem Reviews

A postlaunch review is an opportunity for the usability assessment team, along with the product development team, to discuss and document any outstanding issues that remain when the product is launched. Although it sounds like a housekeeping exercise, it's important to get this information before the teams are disbanded or set upon other projects, so outstanding issues or significant difficulties encountered during the development cycle that remain unresolved can be addressed in the next version of the product.

Following each version of the DSL self-installation kit that was launched, the product team would gather and make notes about usability (and technical) problems that weren't solved before the kit was released. Support groups would bring early results from the help desk describing issues that might not have been captured during the usability assessment. This enabled the usability group to assess why these issues might not have been captured and to refine methodologies to ensure greater success in following efforts. In this way, continuity of effort was maintained and the lessons learned from previous efforts were captured and passed along to the next team members.

A postmortem review, on the other hand, is an assessment of a product after it has been removed from the marketplace. These reviews tend to be less common because any failings with a product that is no longer offered for sale are often seen as "water under the bridge." However, it can be extremely important to understand why a particular product failed or was pulled from the market and what lessons can be learned in the fielding of similar (and, it's hoped, better) products in the future. The key feature of a good postmortem review is that it is not an effort to assign blame but, rather, an effort to lay the foundation for improved products in the future.

This chapter has outlined everything you need to know to perform a basic usability assessment. The next section describes some situations you may encounter when everything doesn't go exactly as planned. Chapter 6 includes some special cases of assessment in which deviations from this method might be required. Chapters 7 and 8 present two case studies to help you more fully understand how usability assessment methods are applied to a variety of products and services in the real world; Chapter 7 describes a formative assessment of a developing Web site, and Chapter 8 details a summative assessment of a state-of-the-art voting machine.

5F) When Things Go Wrong

The process of preparing for and conducting a usability assessment has been portrayed as fairly linear, with the assumption that if you prepare sufficiently, everything will go as planned. In this book I have included a number of tips about how to plan and prepare to ensure that you *don't* get into trouble. That said, there are times when things go wrong, and this section provides additional advice for practitioners who may find themselves in these more advanced situations.

Table 9: Elements That Should Be Included in a Report That Conforms to the Common Industry Format for Summative Usability Reports

Title page

Executive summary

Introduction

- Full description of the product/service/system
- Test objectives

Method

- Participants (demographics, numbers)
- Context of the test
    - Description of the tasks used
    - Description of the facilities used
    - Description of the experimenter's equipment and data collection mechanisms
- Experimental design
    - The procedures used
    - General instructions given to the participants (including IRB and debrief)
    - Specific task instructions given to the participants
- Usability metrics
    - Effectiveness metrics used
    - Efficiency metrics used
    - Satisfaction metrics used
    - Any other metrics collected during the test

Results

- Data analysis
    - Data collection methods
    - Data scoring methods
    - Description of the statistical analysis techniques used
- Presentation of the results
- Tables and graphs with performance data

Note: The CIF does not include a section for recommendations.

5Fi) You Can't Run an Assessment With Users

Sometimes there's not enough time to perform a full usability assessment, no money was budgeted for human participants, or upper management isn't convinced that the time and expense is even worth it for this particular product. What's a usability professional to do?

At a minimum, you can perform heuristic assessments and cognitive walk-throughs on the product or service. This is unlikely to give you the same quality results that are obtained with real users, but these techniques will likely identify products that have significant usability issues and enable you to alert your management of a product that was likely to fail. With this information, you might be able to convince them of the immediate benefit of performing at least small-scale user testing.

5Fii) You Can Run Only a Few Participants

If you have access to a very small number of participants due to time or budgetary constraints, make sure that you maximize the data obtained from your limited testing group. There are two distinct strategies to approaching this problem. If you're testing consumer products, the first strategy suggests focusing on the most frequent tasks that users will be performing with your product. In this way, you'll have a higher probability of catching issues for the tasks that most of your users will be attempting.

If you're building mission-critical systems (military or space systems, for example), the second strategy suggests that you use your limited subject pool to examine those scenarios in which life or limb may be endangered by usability issues. These scenarios may have been identified in a safety review, but they are often found during the selection of tasks for the cognitive walk-through.

5Fiii) The Test Plan Isn't Working Out, and You're in the Middle of Testing

This happens to even the most experienced practitioner. Even though the test plan was meticulously detailed and piloted, an issue arises partway into the test that clearly calls the results of the assessment to date into question. Perhaps the product you're testing relies on wireless Internet access, and although the network was working well when you did your pilot, it's not working well *now*, when you've just tested a dozen subjects. Or perhaps your initial marketing profile of the user was off the mark, and all of the users that you had in the lab didn't even understand the concept of the system, let alone have the skills to use it. Although these details should have been caught in pilot testing, through the luck of the draw, all of your pilot participants had additional skills that allowed them to successfully use the system.

If you're suffering a technical problem, the best you can do is to solve it and move forward. If budget and time allow, discard the tainted data and recruit more subjects to meet your original target. If you can't, however, fall back to the strategies described earlier when dealing with small numbers of participants. In some cases, the data you have may provide excellent information to the design teams, but the details of the test plan will need to be modified to get the information that you were originally seeking. The main advice in this case is to fix what you can and keep collecting data. As the famous English mathematician Charles Babbage once noted, "Errors using inadequate data are much less than those using no data at all."

5Fiv) The Product Failed Usability Testing, But Your Company Is Launching Anyway

Almost every usability professional will encounter this unfortunate situation if he or she has been conducting tests in the real world long enough. Most of the time it's "just business," and executives have had to make a difficult decision about how to best serve the needs of the company. If you're building an e-commerce site and the failures will cause your potential customers to be unable to purchase items, your best strategy is to ply your social skills with decision makers in marketing and product development to determine how usability fixes can be most quickly deployed.

On the other hand, if usability failures are egregious and involve the potential for loss of life or limb, more drastic measures may be necessary. For example, if your usability testing showed that 90% of the users committed an error with a medical device that would result in the death of the patient, immediate discussions with senior company executives are warranted. In the most extreme situations, when loss of life is imminent, whistleblower protections may need to be utilized. Always try to be a team player, but if human life is involved, let your conscience (and a good lawyer) be your guide.

5G) Suggested Reading

- *Handbook of Usability Testing*, by Jeffery Rubin and Dana Chisnell. John Wiley & Sons, 2008.
- "Quantifying usability: The Industry Usability Reporting Project," by Jean Scholtz, Anna Wichansky, Keith Butler, Emile Morse, and Sharon Laskowski. In *Proceedings of the Human Factors and Ergonomics Society 46th Annual Meeting* (pp. 1930–1934). Human Factors and Ergonomics Society, 2002.
- "How to specify the participant group size for usability studies: A practitioner's guide," by Ritch Macefield. *Journal of Usability Studies, 5*(1), 34–45, 2009.
- *Quantifying the User Experience: Practical Statistics for User Research,* by Jeff Sauro and James Lewis. Elsevier, 2012.

CHAPTER 6: SPECIAL CASES OF USABILITY ASSESSMENT

6A) Usability Testing With Telemetry

In Chapter 4, section 4B, I described methods for remote usability assessment that required the involvement of the experimenter. In today's big-data world, some assessments of usability can be made by evaluating the telemetry provided by the technology or product itself. This telemetry also can be used in situations when the experimenter is interacting with the user, by providing tasks or recording during a session. However, the method described in this section focuses on the collection of data from large numbers of users who are engaged in normal activities with the product.

One of the most common applications of this technique is the assessment of Web page usability through log analysis. Through a careful analysis of server-side log data, it's possible to gain an understanding of how users are navigating through the site and what kinds of difficulties they may be encountering. Telemetric data can be simple, using metrics such as time on page or clicks on a specific link, or they can be made up of more complex elements, such as mouse movements (Arroyo, Selker, & Wei, 2006) or navigation paths through a site (Cooley, Mobasher, & Srivastava, 1997).

For example, if the majority of your users are engaging in a multistep process on a Web site (such as putting something in a cart and then ordering it), but the data indicate a sharp drop-off at a specific page, this might signify some usability difficulty with that page. Other kinds of user behaviors that might indicate problems include cycling behaviors, where users move rapidly back and forth between multiple pages and the selection of numerous, if not all, of the navigation elements.

Both of these behaviors can indicate that a user is looking for information but having significant difficulty. This can also be inferred from multiple keyword searches using slightly different search terms each time. Again, the user is probably looking for a specific item or piece of information and is unable to find it.

One of the major difficulties with this big-data approach is how to interpret the telemetry. Because the experimenter does not have direct control over the tasks that users are performing and is not observing the completion of a given task, he or she can only make *inferences* about specific user behaviors based on the data. Often these inferences are correct, particularly if they are averaged over large numbers of users and the behavior is repeated again and again. Hence, analysis of single-user transactions can be more difficult to interpret.

For example, if the telemetry data show that one page in the multistep process is viewed significantly longer than all the other pages in the process, and this observation is based on 100,000 users, the inference that something might be wrong with the page is probably correct. If, however, the inference is based on a single user's telemetry, that conclusion is much riskier. Perhaps the user needed a coffee break or was distracted by a colleague. It could be that the user was struggling with that specific page, but there is no way, from a small number of logs, to differentiate a usability issue from these other potential causes.

Assessing user behaviors over large numbers of transactions with high numbers of users can increase your confidence in results obtained from telemetry. Some companies perform experiments in usability by releasing multiple versions of a page that is involved in some transaction, randomly assigning one of those versions to users as they visit the site, and then observing (via telemetry) which of the versions results in the greatest effectiveness and efficiency. This kind of testing is often referred to as

A/B testing (Chopra, 2010). For high-volume sites, these kinds of methods can be extremely efficient and cost-effective to implement.

Analyzing usability in this fashion has one important benefit. The users are interacting with the Web page in the most naturalistic way possible, on equipment of their choosing and at a time and place of their choosing, so the fidelity of the user actions is high. When used in conjunction with standard usability assessment techniques, particularly as part of the postlaunch review described in section 5E, assessing usability via telemetry can be a powerful tool, but it may not provide optimum results if it is used alone.

This technique, of course, is not limited to Web pages. Tossell and his colleagues (2012) demonstrated the efficacy of this technique in their assessment of smart phone interfaces using specially instrumented iPhones that captured all the telemetry available on the telephone (e.g., weblogs, accelerometer data, location data, e-mail and chat logs). This kind of telemetry is becoming more common, and behavioral data are being collected on a wide variety of products, from digital video recorders (Darnell, 2007) to automobiles (Olson, 2014).

6B) Medical Device Testing

The usability testing of medical devices tends to be more important than usability of other consumer-grade products. If your company builds a clock radio and it's difficult to change the radio tuning settings, the result is that the user might not be able to listen to his favorite station. If you're designing an epinephrine injection pen and people can't use it, the result can be the potential loss of life. For this reason, the testing of medical devices must be done with extra care.

Recall the example in Chapter 2, when hundreds of patients were exposed to significant radiation overdoses due to the usability failure of an imaging system. Indeed, much of the focus on the necessity for good usability practices in medical device design came about after the Institute of Medicine published its seminal report, *To Err is Human* (Kohn, Corrigan, & Donaldson, 2000). That report stated that more than 98,000 deaths each year were attributed to preventable medical errors, many of them usability related. This report served as a call to arms for the human factors and usability community and highlighted the need for rigorous usability and human factors practices in medical equipment design and assessment. Still, recent research suggests that we have not made substantial progress in reducing these deaths, as preventable medical error is now reported to be responsible for up to 400,000 deaths per year in the United States alone (James, 2013).

One of the big differences between medical products and other consumer products is that if you are developing a medical product, federal law, as administered by the Food and Drug Administration (FDA), *mandates* usability verification and validation of that device as specified in ISO/IEC 62366 and ISO/IEC 60601 (International Organization for Standardization, 2006, 2007, respectively). The FDA has issued draft guidance on conducting summative usability assessments on medical products as part of its approval process (U.S. FDA, 2011).

This draft guidance document describes the best practices in defining users, environments, and interfaces, as well as methods of evaluation, including formative and summative evaluation (which the FDA calls *validation testing*). In many respects, the recommended output from this guidance looks very much like the Common Industry Format guidelines. Because of the legal complexities involved, if you're going to be testing medical devices as part of a formal development or approval process, it is imperative that you employ an experienced usability professional in this area.

It is important to separate the legal requirements to perform FDA-mandated summative usability testing from other medical device usability assessments, because not all usability testing of medical devices is conducted to satisfy FDA requirements. Hospitals, independent research groups, and consumer groups may perform usability assessments to obtain information about the usability of devices to make purchasing decisions, to increase the effectiveness and efficiency of their organization, or in furtherance of a specific research goal. For example, numerous researchers (e.g., Carroll, Marrero, & Downs, 2007; Gao & Kortum, 2015; Surabattula, Harvey, Aghazadeh, Rood, & Darisipudi, 2009) have performed usability assessments on home health care devices where adherence to FDA guidelines was not required.

There is an important difference between the usability assessment of medical devices and that of consumer goods. Medical devices often require invasive or potentially harmful procedures. We tend to think of complex surgical equipment, such as the da Vinci surgical robot or a ventilator, as the kinds of devices that are invasive and potentially dangerous, but simpler devices that are often used by health-care professionals and consumers alike also fall into this category (e.g., blood glucose monitors, diabetes injection systems). Clearly, assessment of these kinds of systems requires the involvement of clinical personnel and approval by the participating medical institutions or similar Institutional Review Boards.

During early formative stages, the use of simulators or facsimile devices that do not require invasive procedures and pose no danger to the participant can be used. For example, if you're testing an automated electronic defibrillator, it wouldn't be necessary to provide the final electric shock, as this could be simulated, in which case the device would pose no danger to the participant or the test patient. An example of a patient simulator that could be used in a usability assessment is shown in Figure 12.

Wiklund, Kendler, and Strochlic (2010) provided excellent guidance for the usability testing of medical devices, and readers who will be performing assessments in this specialized domain would be well advised to consult that book.

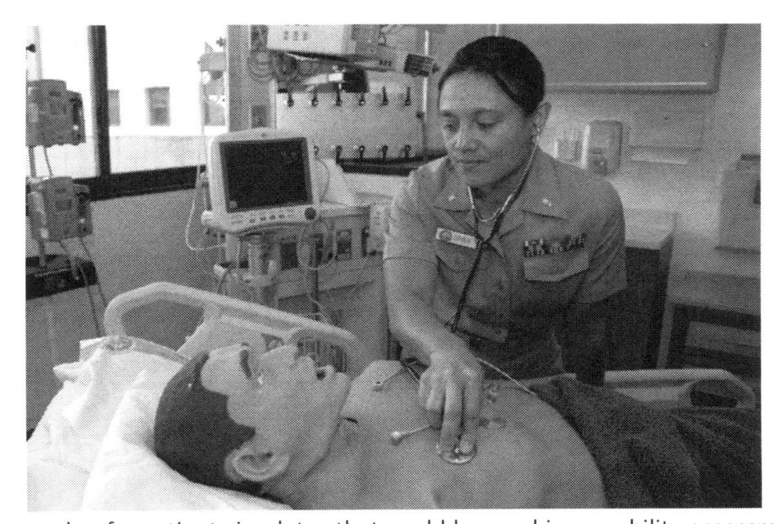

Figure 12: An example of a patient simulator that could be used in a usability assessment (*public domain image courtesy of U.S. Navy*).

6C) Mobile Device Testing

A number of unique issues are associated with the usability assessment of mobile devices – for example, mobile phones. Most notable among these is the fact that it can be difficult for the experimenter to observe what the user is doing, given the small format of these kinds of devices and their displays and the way the devices are handled when in use.

There are several possible ways around these difficulties. The first involves use of a fixture that holds the phone in a specific position so the interaction can be observed by a camera above the phone. In this way, the experimenter can see, for the most part, what the user is doing. Although this is certainly a reasonable solution for understanding fundamental user experience flows and interactions, it tends to result in a degree of artificiality, because users don't hold and work with their phones in this manner. Further, users often obstruct the camera as they interact with the phone, so a perfect record of the interaction is not always captured.

An example of a camera-based fixture is shown in Figure 13A. This device occlusion can be mitigated by doing an electronic screen capture, but then you don't get information about how the user is physically interacting with the device. Combinations of screen capture software and physical observation cameras have proven to be reasonable solutions.

Another potential solution is the use of a cell phone emulator. In this case, a facsimile of the cell phone is programmed into the computer, and that functional image is displayed on a standard computer monitor. Using the mouse and keyboard, the user can then interact with "phone" on any number of tasks. The advantage is that the standard methods of observing computer screens can be utilized. The disadvantage is that users' physical interactions with the device are completely artificial. An example of a smartphone emulator is shown in Figure 13B.

In the end, you'll need to decide which of these methods provides the best information, given the kinds of tasks you'll be doing and the context in which those tasks are normally performed.

Another difficulty associated with testing the usability of mobile devices is that they are highly context dependent. Certainly it can be argued that all devices being tested are context dependent. However, when testing a product such as a printer or can opener, you can be reasonably certain how and where the device will be used in the field. With mobile devices, on the other hand, the environment in which the device and its associated software will be used is nearly limitless. Further, these different environments can have a significant impact on the ability of users to perform specific kinds of tasks.

For example, using a mobile device on a crowded train as you're being jostled back and forth, or when you're walking down the street, can add significant difficulties for users trying to perform a number of tasks. Some of these environmental factors can be simulated in the laboratory (Kjeldskov & Stage, 2004), but field studies bring an extra degree of fidelity. Field studies also introduce additional issues in interacting with the user and recording actions that the user undertakes as he or she completes tasks (Zhang & Adipat, 2005).

Some researchers have shown that results obtained in the field and laboratory are relatively equivalent (Kaikkonen, Kallio, Kelalainen, Kankainen, & Canker, 2005), whereas others have found that field studies produce richer data (Duh, Tan, & Chen, 2006). These differences may be due to the context dependency of the tasks that were performed. For example, tasks in which there are significant environmental factors may benefit from the use of field studies. The use of simulators to establish context for mobile usability in situations that may be dangerous is highly recommended. For example, there are numerous studies on the usability of cell phones while driving (e.g., Serafin, Wen, Paelke, & Green, 1993; a review by Baron & Green, 2006), and this kind of study would be difficult – if not impossible – to ethically perform in the field. In this case, driving simulators have been employed so users can safely assess the usability of phones while performing the primary task of driving. An example of such a simulator is shown in Figure 14.

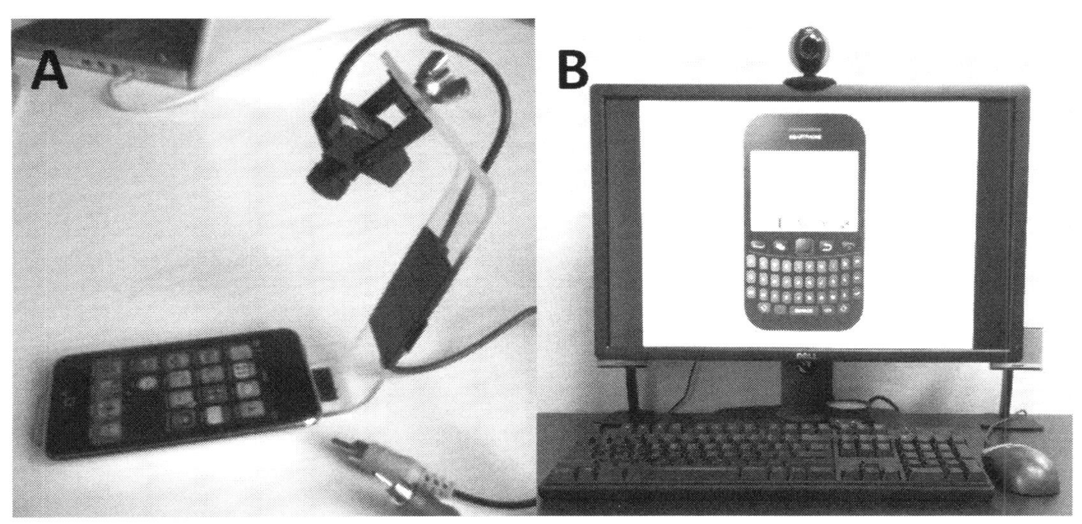

Figure 13: **(A)** An example of a camera-based fixture (*image by Nick Bowmast, used with permission*), and **(B)** an example of a smart phone emulator (*copyright by Philip Kortum*).

Figure 14: An example of the usability of a cell phone being assessed safely in a medium-fidelity driving simulator (*image USMC, public domain*).

One way around many of the difficulties associated with performing usability assessments of mobile devices described earlier is through the use of customized logging software, in which the actions of the users are recorded in the background asynchronously for review by the experimenter at a later date. The logging software is essentially invisible to the user and records all pertinent elements of the user's interaction. In this way, the user can engage in highly realistic behaviors in the real world without intervention by the experimenter. Further, because there is no need for intervention, this type of methodology supports longitudinal assessments of usability, and the resulting log files can be uploaded periodically. This method has this been described for use in iOS devices (Tossell, Kortum, Shepard, Rahmati, & Zhong, 2012) and Android devices (Shabtai & Elovici, 2010).

Finally, much of the utility of mobile devices centers on their ability to send and receive data. For this reason, the strength of the data connection can have a significant impact on people's ability people to perform tasks. More important, unless you're using the automated logging tools described earlier, the degree and severity of these fluctuations in available connectivity are unknown and can have uneven impacts on user performance.

6D) Subjective Assessment Without Specific Tasks

In certain special cases, it may be advantageous to collect subjective usability assessment data from users without conducting specific task-based usability tests. A subjective usability measurement instrument, such as the System Usability Scale, enables users to provide data on the usability of a product, service, or system with which they are familiar but without performing specific tasks under the direction of an experimenter (Kortum & Sorber, 2015; Kortum & Bangor, 2013, Kortum & Peres, 2015). Applying this method is straightforward: The user is asked to think about the entirety of his or her experience with a given product or system and then to rate the usability of that product using a subjective usability measurement instrument (e.g., the SUS).

The advantage of this kind of assessment is that it can produce a more complete picture of the usability of a complex or multifunction device. Users can make their usability assessment based on the totality of their experience of use with the product, rather than on the basis of their interaction within the context of a few tasks.

Some researchers (Kortum & Bangor, 2013) have demonstrated the utility of this kind of assessment in measuring the usability of *classes* of products (e.g., microwaves) as well as the assessment of specific products (e.g., Acme microwave model X123). Assessing the usability of classes of devices has utility if you're trying to compare your product or service with dissimilar products and services. It also can be of value when trying to answer questions like "Should I deliver my help services using the Web, the telephone, or messaging?"

This kind of retrospective assessment is not without its pitfalls. Users may have had a good or bad experience with the product recently, and that experience could bias their assessment. Further, although integration of the assessment across tasks may be viewed as a strength, it is unclear which elements of the interface are actually being evaluated. Some users may exercise only one or two of the common core functions, whereas others may be utilizing difficult advanced features, and the tasks users are rating are typically not captured in the data. One potential remedy in such cases is to increase the sample size to ensure that you have confidence in the average values that are obtained. Fortunately, the method is extremely cost-efficient and can be administered remotely without experimenter intervention, so it can be practical to measure a large number of users.

Don't be tempted to use this form of assessment for products and services in your development cycle. Although it is extremely cost-efficient, it does not capture any of the information about the kinds of difficulties that users are having or the frequency with which they're having them. It provides only a single snapshot score for a given product, which can be misleading if the data are not collected correctly and interpreted with the limitations of the method in mind.

6E) Special Populations and Populations With Disabilities

If your product or service is targeted for the general population (e.g., cell phones, microwave ovens, productivity software) and not a highly specialized population such as astronauts or fighter pilots, you'll need to consider including persons with disabilities in your test plan. In many ways, these users are no different from your able-bodied users, in that they fit the user demographic for your product (income, education, gender, etc.) and want to purchase and use your product. If your product design team has done its job, the product should be capable of serving a highly diverse demographic, including persons with disabilities.

Unfortunately, even if your design team has rigorously followed the best design practices for populations with disabilities (e.g., Web Content Accessibility Guidelines, or WCAG), fewer than half of accessibility-based usability issues will have been designed out (Rømen & Svanæs, 2012). In some cases, you may recruit specifically to include populations with disabilities to capture all the relevant information. In other cases, if you recruit from the general population, users with special needs may simply be part of your recruited population and you won't realize it until they show up for your test.

One difficulty in testing persons with disabilities is sampling broadly enough to capture the range of visual, hearing, and physical disabilities that are prevalent to generate valid results. Although budget constraints may limit your sample sizes, it is suggested that you recruit as representative a sample across the range of disabilities that you can (van der Geest, 2006).

One unique characteristic of users with disabilities is that they often utilize specialized equipment to help them access and use certain technologies. For example, users who are blind or visually impaired frequently have specialized screen reader programs or dynamic Braille devices that help them to "see" a Web page. If you bring users into your laboratory and don't have that specialized equipment, it is impossible for these users to use your Web site. More important, we have found that these types of users typically customize those accessibility devices to a significant degree, so even if you have the required hardware and software, having your subjects use it can be problematic.

For this reason, it is often advantageous to go to the user's site (as described in section 4Bii) to enable users to access all the technology they would normally need to perform a given task. As noted earlier, schools for the blind and visually impaired and schools for the deaf and hearing impaired are excellent places to recruit these kinds of users. Testing remotely is also an option, but research has shown that although quantitative results are comparable to those found in usability labs, qualitative results tend to be less robust (Petrie, Hamilton, King, & Pavan, 2006).

Do I really have to test with special populations?

Given that 12.1% of the U.S. population is known to have some form of physical or cognitive disability (Erickson, Lee, & von Schrader, 2014) – that's approximately 37 million people in the United States – it seems to be a good practice to test these individuals, as they are likely to be in your user population. More important, if you're building software, Web pages, multimedia products, computers, or telecommunications products that will be used or procured by federal and state governments, the law requires that they be accessible by persons with disabilities, and testing is the only way to ensure that these populations can use the product. The accessibility of these products is governed by Section 508 of the Rehabilitation Act and Section 255 of the Telecommunications Act. Design teams should carefully consult these documents to gain an understanding of what constitutes accessibility. Jaeger (2006) provided an excellent guide on how to assess compliance with these federal regulations.

The relationship between usability and accessibility is complicated. Certain features and functions that make products and services more accessible for users with disabilities can also enhance usability for users who do not have a disability (e.g., closed-captioned text for television). Other accommodations may not add to the usability of a product for a user who is not disabled, but they can add significantly to the ability of a user who is disabled to operate the product or service (e.g., text-to-speech conversion of Web pages). In general, all major functions of any product or service should be tested for usability, and this includes features for accessibility.

Elderly users who may be suffering from palsy, hearing and vision deficits, and cognitive deficits also fall into this category of special populations. Many of those users do not have specialized equipment and expect the equipment that you are providing will support them in their task completion. In these cases, the usability mettle of your product will be significantly tested. Recruiting elderly users can be somewhat easier, as community day centers and retirement homes can provide a large supply of such users.

6F) Suggested Reading

- *Usability Testing of Medical Devices,* by Michael Wiklund, Jonathan Kendler, and Allison Yale Strochlic. CRC Press, 2016.
- "A meta-analytical review of empirical mobile usability studies," by Constantinos Coursaris and Dan J. Kim. *Journal of Usability Studies, 6*(3), 117–171, 2011.
- *Computer Systems Experiences of Users with and without Disabilities: An Evaluation Guide for Professionals,* by Simone Borsci, Masaaki Kurosu, Stefano Federici, and Maria Mele. CRC Press, 2013.
- "Usability ratings for everyday products measured with the system usability scale," by Philip Kortum and Aaron Bangor. *International Journal of Human-Computer Interaction, 29*(2), 67–76, 2009.
- "Assessing Section 508 compliance on federal e-government web sites: A multi-method, user-centered evaluation of accessibility for persons with disabilities," by Paul Jaeger. *Government Information Quarterly, 23*(2), 169–190, 2006.

CHAPTER 7: REAL-LIFE EXAMPLE 1: FORMATIVE ASSESSMENT OF A CORPORATE WEB PORTAL

Now that you've read about all of the steps necessary to complete a rigorous usability assessment, Chapters 7 and 8 will walk you through two examples of these methods in detail. Recall from Chapter 3 that there are two fundamental forms of usability assessment: The *formative assessment*, in which the goal is to provide input to an ongoing development process, and the *summative assessment*, in which the goal is to establish a usability benchmark against which other products or other versions of the product can be compared. Chapter 7 describes a formative usability assessment of a corporate Web portal, and Chapter 8 describes a summative usability assessment of a high-security voting system.

This first case study describes the formative usability assessment of a Fortune 100 telecommunications company's main Web site. In today's Internet-connected world, having a well-designed corporate site is essential to the functioning of a business. Customers (and potential customers) use such sites to gather information, obtain help, pay bills, and make purchases. Well-designed sites can also generate significant cost savings for companies if the sites enable users to perform functions that would otherwise require the customer to interact with a live human service representative. Because of the great importance in having a Web site that is easy to use, companies devote significant resources to ensuring that their sites contain as much usable functionality as possible.

As with most telecommunications firms, this company had several lines of business – including landline telephones, mobile phones, business phones, and Internet service – spread across several regional brands. Prior to the redesign, each of these lines of business and brands had its own separate Web site. As you can imagine, this did not create an integrated experience for users who needed to interact with more than one of these businesses.

A decision was made by the corporate leaders to establish a single unified and integrated Web presence for all the company's different lines of business and brands. A professional Web development firm was brought in to establish the design, and the corporate human factors group was enlisted to aid in the development to ensure that the resulting site was highly usable in addition to being aesthetically pleasing.

The Web development firm created fully interactive prototypes running on a private and protected server so development and usability assessment could take place independent of the day-to-day operations of the Web sites that were in use. Because this was part of a development effort, we conducted a series of formative assessments to aid in the design of the integrated Web site. This case study is a composite of several formative assessments that were performed over several months.

7A) Define the Purpose of the Test

The purpose of the formative assessment was to provide specific input to the professional Web developers to aid them in identifying and mitigating any difficulties that users encountered during testing of the integrated site. We also wanted to establish usability benchmarks for the new integrated Web site to determine if it could perform at least as well as each of the individual sites it was replacing. Because this was an iterative process, the benchmarks would also serve as a way to ensure that the site was getting better from version to version.

7B) Define the Users

Demographic information was collected by the marketing department for each of the separate sites. We had hoped to develop a single integrated user profile to help define our users. However, examination of the data indicated that two separate user profiles needed to be accommodated during testing. The users of the landline phone site, the mobile phone site, and the Internet site all had similar profiles and were combined into a single "consumer" user profile, whereas business telephony users remained a separate profile. For simplicity, we will focus on the formative tests conducted for the consumer side of the Web site.

Because of the ubiquity of telephone (landline and mobile combined) and Internet use, these users comprised a wide range of demographics. Marketing data indicated that the user population for this test was essentially everyone over the age of 18. Although some product lines (like Internet) assumed higher education, socioeconomic status, and technical skill levels (e.g., familiarity with computers), other product lines (e.g., landline telephones) were utilized across all education and socioeconomic levels. For this reason, we did not generate any requirements regarding the economic status, skills, prior knowledge, or abilities, except that we required that users be existing customers of the company at the time of testing, or someone who was looking to become a customer. Using this definition, we were able to capture people who were most likely to be users of the site while still matching the very broad demographic suggested by the marketing data.

7C) Perform a Preliminary Assessment I: Heuristic Assessment

To prepare for the heuristic evaluation, we asked the Web development firm to provide us with their first stable beta version of the Web site. It is important to note that we did not wait for the final version of the site to become available, which provided us an opportunity to fix any issues identified by the heuristic assessment before we brought in real users.

A single heuristic assessment was performed for the integrated site, and we used the general-purpose heuristics that were defined earlier in Table 3. Because these heuristics were developed specifically for Web pages, no system-specific heuristics were added. Issues resulting from the heuristic evaluation were exceptionally lengthy, as is typically expected in early formative assessments. The output of the entire heuristic is not included here, but sample outputs are shown in Table 10.

7D) Define the Tasks

For this test, we defined a broad range of tasks intended to exercise the site to the greatest degree possible within the 50 minutes allotted for each user test, assuming a 5-minute limit for each task. The tasks were developed by reviewing both sales and repair call-center logs and then selecting questions that were frequently asked by our customers (e.g., Tasks 1, 3). Other tasks were developed based on inputs from the marketing department on activities that were deemed of potentially high value for our customers using the Web (e.g., Tasks 2, 8).

Table 10: Selected Outputs From the Heuristic Assessment for the Integrated Web Site

Heuristic	**Usability Deficiency That Violates the Heuristic**
Speak the user's language.	The site combines language for consumer, business, and investors all on the home page, making portions of the site difficult for some user groups to understand.
Minimize the user's memory load.	The site is extremely cluttered, forcing the user to perform long serial searches for links and then remember where those links were located.
The interface should have consistency.	Some of the links go to pages that still have the old branding on them, making it difficult for users to determine their location.
Sufficient feedback should be provided.	Some links lead to noncorporate sites, and no feedback is provided to the user that they are actually leaving the company's site.
Clearly marked exits should exist.	Once a user clicks on a link on the home page, there is no way to get back to the home page without using the back button provided in the browser.
Shortcuts should be available.	No shortcut is provided to the online shopping area, which forces users to navigate through multiple pages to get to this important function.
Precise and constructive error messages should be used.	On the ordering Web page, if a required field is left blank, the page simply reloads, giving no indication of what, if anything, is wrong.
Help and documentation should be available.	Help documentation is provided only for consumer services; none is available for business or investor functions.

Task 1: You have just moved to 123 Main St., Tinyville, TX. Order a basic landline telephone for your new home.

Task 2: You currently receive your telephone bill in the mail. Change this so you get it via e-mail instead.

Task 3: Every time you pick up your home telephone, you hear a lot of clicking on the line. Determine how you might fix this problem.

Task 4: You make a lot of long-distance calls, but only to members of your family, who all live in Minnesota. Find the most economical long-distance plan that would support this kind of long-distance calling.

Task 5: Your telephone number is 608-555-1212. Use the Web site to pay your telephone bill this month.

Task 6: What is the current stock price for this company?

Task 7: What is the current price for the Motorola Razr phone if you don't want to get a long-term contract?

Task 8: Determine if you can get high-speed Internet at your address of 123 Main St., Tinyville, TX.

Task 9: You are sending your daughter off to college this fall at West Central University. Order phone service for her dorm room.

Task 10: You have been getting a lot of unauthorized telemarketing phone calls from 501-555-1212. Determine who is making these calls so you can report them to the authorities.

Although several of the tasks – notably 1, 8, and 9 – appear to be redundant, they actually require the user to exercise distinctly different portions of the site, and this is why they were all included. Also note that these tasks were given in random order to minimize the impact of the sequence of test tasks on a user's ability to do the task.

7E) Perform a Preliminary Assessment II: Cognitive Walk-Through

Using the tasks just defined, we performed a cognitive walk-through on the integrated site. Because this was a formative assessment, we waited until all of the issues that were identified in the heuristic evaluation were fixed before we proceeded with the cognitive walk-through. This ensured that we weren't making evaluations of issues that we had already identified in the previous heuristic assessment.

We used the streamlined cognitive walk-through method described by Spencer (2000), detailed in section 3F. Let's examine a portion of the output of the cognitive walk-through for Task 9, in which the participant is asked to order phone service for her daughter's dorm room at West Central University.

1) *Will users know what to do at this step?*
   a) More than 50 active links are available on the main page, which will require the user to do a serial search to find the relevant link.
   b) The most prominent link on the homepage that is a potential target is labeled "order new phone service," but this link will not work for ordering college phone service activation.
   c) Users must go to the "residential customers" page to find the correct link, and this label may be insufficient to suggest to users that this is the correct link to follow.
2) *If the user does the right thing, will he/she know that he/she did the right thing?*
   a) Because the actual ordering page for college phone service is still 5 clicks away, the user may not have a good indication that following the "residential customers" link was the correct action.
3) 3, *USER IS NOW ON THE "RESIDENTIAL CUSTOMERS" PAGE. Will users know what to do at this step?*
   a) More than 30 active links are available on this page, which will require the user to do a serial search to find the relevant link.
   b) On this secondary page, the link that takes users to the college telephone service signup page is not colored or underlined like the rest of the links on the page, so users may not know that it is clickable.
4) *If the user does the right thing, will he/she know that he/she did the right thing?*
   a) The actual ordering page for college is still 4 clicks away, but the page is labeled "college options," so the user will have a good idea that he/she is making progress toward the goal.

The cognitive walk-through clearly identified that this was likely to be a difficult task for users to perform. They would probably suffer significant confusion in trying to determine what to do at each of the individual steps; more important, even if they did the right thing, it may not have been evident that they were making progress toward their goal. Worse, if they did the *wrong* thing, the system gave them the same feedback as if they had done the right thing.

The cognitive walk-through for this task suggested that significant redesign would be necessary to make completing this task easier. Subsequent testing with real users confirmed that this particular task was exceedingly difficult for users to complete.

7F) Create the Test Plan

The test plan for the Web site included all of the necessary elements required to conduct the usability assessment. Each of these elements is detailed next.

7Fi) Define the Metrics

We used the three metrics of effectiveness, efficiency, and satisfaction, as defined by ISO 9241–11. These metrics were collected for all 10 tasks. Task effectiveness was determined by whether a user was able to successfully complete the task. Efficiency was determined by the overall time it took a user to complete the task, the number of times he or she clicked a link, and the total number of clicks made while performing the task (to account for users repeatedly visiting the same link). Satisfaction was measured using the After Scenario Questionnaire (ASQ) and the System Usability Scale (SUS).

At the time, it was assumed that customers would probably come to the Web site to perform a single task and do so infrequently (call centers were still seen as the primary way to interact with the customer), so none of the secondary metrics of usability (learnability, workload, likelihood to recommend) were used during the study.

7Fii) Define the Testing Environment

The test was performed locally in a state-of-the-art, multiroom laboratory facility. As can be seen in Figure 15A, the user was seated in front of the computer in a home office environment and performed the task while alone in the room. The room was fully instrumented to enable audio and video recordings as well as a full-screen capture of the user's monitor. Three separate cameras captured any view of the user that might have been desired. Software that counted the number of clicks and mouse movements was installed on the computer for additional data collection.

Figure 15B shows the control room for the test facility. The video recording was able to capture three simultaneous image streams and combine those into a single recordable signal that was captured on tape. The cameras were fully adjustable and completely silent so they could be readjusted to capture new elements of interest in the user's room. Note the panel on the left, which allowed us to supply any resource (telephone line, DSL, ISDN, Ethernet etc.) into any port in any user room, making it easy to configure the laboratory for different kinds of tests.

The experimenter sat in front of the control panel, and a small number of observers were allowed to sit directly behind the experimenter so they could easily see the participant through the one-way glass. When there were more observers than the space would support, the video signal could be sent to an adjacent conference room to accommodate this need.

Timing and data logging were performed with a commercially available software package that was integrated with the video streams for easier data analysis. A second experimenter was always present to record notes on the user's actions. This facility could support four simultaneous usability tests.

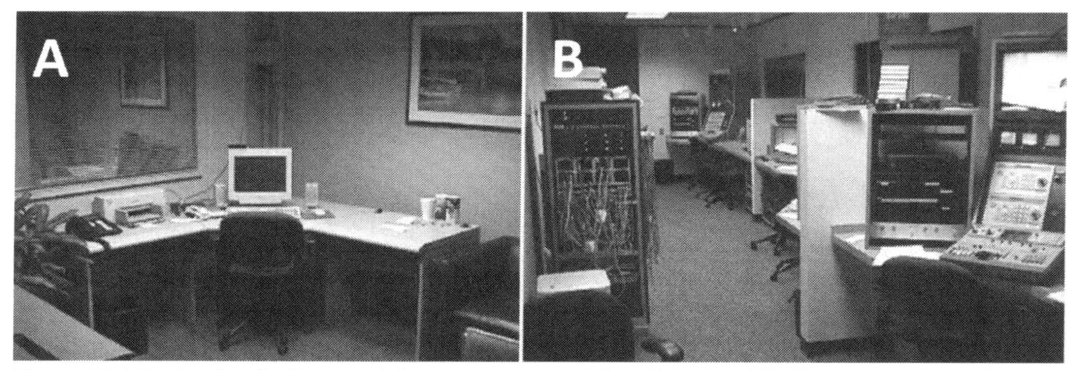

Figure 15: The testing facility used for assessment of the Web site. **(A)** The room where the user performed the test and **(B)** the control room. Note that the laboratory is fully instrumented and has a one-way mirror for observation (*image copyright by Philip Kortum*).

7Fiii) Define the Timing Parameters

Timing was performed using the logging software that was integrated with the audio and video recording equipment. Start and stop times were recorded by pressing a specific key on the keyboard. Each task had its own time measure; overall time for the 10 tasks was not recorded.

For each task, the computer was set to the home page, and the screen was blanked by the experimenter. The participant was given the task in writing, and, after indicating that he or she was ready to begin, the task the screen was unblanked and time recording was started. Timing was stopped when the user stated, "I am finished," as instructed to do at the beginning of the task. In some cases, users failed to make the finishing statement, so the experimenter was able to note when the users made indications (like pushing away from the desk or navigating away from the task Web site) that they were finished.

7Fiv) Define the Testing Materials

7Fiv-1) Product being tested. We used the first stable version of the integrated Web site that was made available by the Web development firm that was doing the coding. The version we used for user testing was the version of the Web site that had been modified based on the output of the heuristic evaluation. Some additional usability issues identified during the cognitive walk-through were also fixed prior to our beginning the user testing, but many of these issues were more global in nature, and it was determined that the first formative assessment should occur to validate the findings.

The Web site was fully complete and operational in every respect, a trait that is not always found in early formative testing. The developers worked with us on building fake back-end systems to support the tasks that we had defined, to enable our participants to use the site to order products, check phone numbers, and make payments without having to access any real back-end systems. For example, the user was required to enter some personal information (address, etc.) to be able to order a telephone line for a new home (Task 1). The Web site would take this information and "validate" it, even though the information wasn't real. Essentially, the development platform would take whatever input it was given and move on to the next step in the process, as if the entered data had been checked and verified in the company's database. This enabled the user to interact more normally with the system.

7Fiv-2) Support material and services. Each of the tasks required specific support material and services to enable the user to perform them in a seamless fashion. Table 11 shows the support material and services that were required for each of the tasks. It's important to provide users with this kind of fake information, because you don't want to have your participants using their own information in these kinds of tests. Many users aren't comfortable doing so and might choose to withdraw from the study, but, more important, it allows you to set up your facsimile back-end systems to accept and recognize the required inputs and then provide appropriate outputs back to the user. Although the telecommunications company had a customer service telephone number, we did not allow this particular support feature to be utilized for any of the testing, so no simulated live representatives needed to be developed.

Note that a significant number of support services needed to be developed to allow the user to conduct a variety of transactions on the Web site. Without the supporting material, the test would not have been nearly as effective because the interactive nature of many of these back-end systems was what drove the usability (or lack thereof) of the Web site.

7Fiv-3) Test materials. We used approved consent and debrief forms for the assessment, along with paper versions of the System Usability Scale and the After Scenario Questionnaire. A demographic questionnaire that inquired about age, gender, computer experience, and telephone services currently owned was also used and administered via paper. Pens, pencils, and a notepad were available so the participants could take notes if they felt it was necessary during the test.

As noted in the facilities section, timing and audio visual data were collected using the data-logging software that was integrated with the audiovisual equipment. Notes about users' actions were captured using a standard laptop computer and word-processing program.

7Fiv-4) Procedures. Total time for all 10 tasks was set at 50 minutes, with a limit of 5 minutes for each task. This limit was based on analysis of historical log data, which suggested that after 5 minutes, real users tended to abandon the tasks. We allotted 30 minutes between participants to reset the browser and the back-end systems and to account for users who might be slightly late.

Because the testing was conducted at a corporate facility, when participants arrived, they were greeted at the security desk, signed in, and given a visitor's badge. Security staff called the experimenter to escort the users to the laboratory. Once the users were seated in the laboratory, we used a a written procedure for administering the informed consent and instructions detailing what the users would be doing and how their tasks would be given to them. Users were told that no assistance could be given and that if they felt they couldn't complete a task, to let the experimenter know by saying, "I can't complete this task." Once users acknowledged that they understood these instructions, the demographic questionnaire was given, and after the questionnaires were completed, the experimenter left the participant's room so the assessment could begin.

Table 11: The Support Material and Services Required for Each of the Tasks

Task	Support Material	Support Service
1. You have just moved to a new home. Order a basic landline telephone for your new home.	• A fake address • A fake previous address • A fake e-mail • A fake previous telephone number	• Database that supports the fake addresses and phone numbers
2. You currently receive your telephone bill in the mail. Change this so you get it via e-mail instead.	• A fake address • A fake telephone number • A fake e-mail	• Database that supports the fake address and e-mail
3. Every time you pick up your home telephone, you hear a lot of clicking on the line. Determine how you might fix this problem.	• A fake address • A fake telephone number • A phone that connects to the fake back-end test system • A phone number for the fake back-end test system	• A fake back-end system that pretends to perform an automated test of the phone line and reports the results
4. You make a lot of long distance calls, but only to members of your family in Minnesota. Find the cheapest long-distance plan.	• A fake telephone number • Fake telephone numbers for the people you call the most	• A fake evaluation system that will look at the numbers most often called and return a suggestion for a long-distance plan
5. Use the Web site to pay your telephone bill this month.	• A fake telephone number • A fake credit card • A fake e-mail	• A back-end system that simulates a payment transaction for given telephone number • A back-end system that will take the fake credit card information and appear to process it
6. What is the current stock price for this company?	• None required	• None required (the information is hard-coded in the site)
7. What is the current price for the Motorola Razr phone if you don't want a long contract?	• None required	• None required
8. Determine if you can get high speed Internet at your new home.	• A fake telephone number • A fake address	• A back-end system that simulates the testing of a line for suitability for high-speed Internet, based on the user's fake phone number and location
9. You are sending your daughter off to college this fall at West Central University. Order phone service for her dorm room.	• The fake college name and location • Fake billing information • A fake credit card	• A database with a list of colleges, including the fake college • A back-end system that simulates the ordering of a new telephone line for college locations
10. You have been getting a lot of unauthorized telemarketing phone calls from 501-555-1212. Determine who is making these calls, to report to the police.	• A fake name of the company that is making the phone calls • A fake telephone number from where the calls are originating	• Database that has the phone call number • Database that can return the name of the fake caller

A written procedure described what to do in case of an equipment failure. If the computer failed, a ready spare was available and would be swapped for the failed piece of equipment. The task that the user was performing at the time of a failure would be discarded, and the user would simply move on to the next task when the replacement computer was ready. Because the tasks used only a standard desktop computer, no emergency interventions were anticipated, although users were informed at the beginning of the test that if there was a fire alarm, they would be escorted from the building.

After the participant completed all 10 tasks, the experimenter returned to the room and asked the user if there were any issues in particular he or she wanted to share about the site. Using procedures developed by the recruiting firm, the participant was then paid by check (after signing a form indicating that he/she had participated in the study and been paid) and escorted back to the security desk for dismissal.

Because observers were allowed (and encouraged) to watch the testing, specific written procedures were developed in a handout so as to make each observer aware of the rules. Specifically, we asked that even though the facility was as soundproof as possible, conversations should be kept to a minimum in case the microphone was on. The procedure further stressed that no help could be offered to the participant and that specific questions for the participant should be held until all 10 tasks were complete, at which time the experimenter would collect the questions and ask them of the user (if they were appropriate).

7G) Recruit the Users

We enlisted 30 users for this formative assessment using a market research firm to identify and recruit the participants. Although this is a rather large number of users for a formative assessment, we wanted to be sure that we were capturing enough data for a more rigorous statistical analysis, given the large scale of the development effort.

In our recruiting specification, we required an equal gender mix and an equal distribution of age within three ranges (18–35, 36–55, 56–unlimited) to ensure broad representation. Recruits also had to be current customers of the telecommunications company, or users who indicated that they were about to become customers. Because this was a test of a Web-based service, users were further required to have Internet access either at home or at work, but we defined no specific minimum limits on their use of the Internet. Smart phones were not deployed at the time of the test, so no considerations were given for this kind of Internet access. No other user characteristics were specified.

The recruiting firm performed all the prescreening and participant interactions prior to their arrival at the assessment facility. They also provided all the materials necessary (checks and legal forms) for the compensation of participants. If a participant did not show up for a test, the recruiting firm would supply a participant of the identical demographic at the end of the test.

7H) Run the Test

Once all of the specified changes suggested by the heuristic evaluation and selected changes from the cognitive walk-through had been implemented in the prototype Web site, along with the completion of all of the necessary fake back-end systems to support the tasks, we were ready to run the test.

7Hi) Pilot

A small pilot of three participants was run to ensure that all of the procedures and materials assembled for the test were functioning as planned. These three participants were supplied by the recruiting firm, one user from each age group with an unspecified gender mix. We adhered to all of the other recruiting criteria.

Because the system had already been thoroughly evaluated and refined following the heuristic evaluation and cognitive walk-through, and the prototype had been rigorously tested for quality assurance by the development firm, no modifications to the procedures or the Web site prototype were indicated during the pilot testing. Even though no additional issues were encountered during the pilot, the data from these initial users were discarded (as recommended in section 5B) before the full test was started, in accordance with the written procedures developed prior to the test. Given the smoothness of the pilot test, the team had no concerns about the readiness of the system or procedures for the full test.

7Hii) Full Test

Participants were scheduled into the laboratory at a rate of four per day, which meant that the entire formative assessment took eight days. After the participants arrived and were checked in by security, they were brought to the laboratory to begin testing. They signed the informed consent, filled out the demographic questionnaire, and were given the instructions on how the test would proceed. The tasks were then given to the participants one at a time, in random order. Participants progressed through each of the 10 tasks until they had completed each one or indicated that they could not make further progress.

Perhaps because the compensation level for the participants was relatively high, only a few participants did not show up for their designated times, and the recruiting firm supplied replacements for completion on the last day of testing. Numerous observers from the Web development firm, marketing department, and corporate office attended the testing over the course of the study to see firsthand how the usability of the site was progressing.

All of the procedures were executed as written, and the full test proceeded in an expedient manner without significant deviations from the plan.

7I) Report the Results

A final report that described the results of the formative usability assessment was written, following the general form of the Common Industry Format. Recall that the CIF is formatted primarily for the output of summative usability assessments. However, the important elements described in the CIF concerning the experimental design, metrics used, and results of the assessment were included. Also included in the report were the results of the heuristic evaluation and the cognitive walk-through, along with descriptions of the changes that were made to the site to fix the deficiencies that were identified in these two pretest assessments.

Because this was a formative assessment, one of the major items in the final report that was required by the development team was a list of usability deficiencies identified in the Web site and suggestions for modifications that would mitigate these deficiencies. Using this list, the development team could then create a revised version of the Web site for additional formative testing. All told, the Web site went through over a dozen formative usability assessments as it progressed through the development process.

Another element in the report that is not identified in the CIF was a description of the severity of each of the errors and a prioritization of the order in which the usability assessment team believed the usability deficiencies should be tackled. The usability team is in a unique position to make assessments about which usability deficiencies may lead to the greatest user difficulties, and so input on which errors should be fixed first can greatly benefit the development team, as time and cost constraints make it difficult to fix all of the errors.

Because of the relatively fast-paced nature of the development process, and the desire of the product development group to question the results and potential fixes, a good portion of the output of the formative test was delivered in a summary presentation format that was presented in person to the product development teams and management. This format of sharing results enabled the developers to question each of the findings in the formal report and receive real-time feedback about potential fixes from the usability team.

7J) Suggested Reading

- "Usability in practice: Formative usability evaluations-evolution and revolution," by Ginny Redish, Randolph Bias, Robert Bailey, Rolf Molich, Joe Dumas, & Jarad Spool. In *CHI '02 Extended Abstracts on Human Factors in Computing Systems,* pp. 885–890. Association for Computing Machinery, 2002.
- *A Practical Guide to Usability Testing,* by Joe Dumas & Ginny Redish. Intellect Books, 1999.

CHAPTER 8: REAL-LIFE EXAMPLE 2: SUMMATIVE ASSESSMENT OF A HIGH-SECURITY VOTING SYSTEM

Voting is the cornerstone of the American democratic system. After the debacle in the Florida presidential election of 2000 (Agresti & Presnell, 2002), voting districts across the country switched to electronic voting machines in an effort to avoid the problems associated with paper and other paper-based forms of voting technology (e.g., punch cards). As these electronic voting machines near the end of their service lives, awareness has increased of the necessity for enhanced security on these kinds of systems. Toward that end, computer security experts from across the world have been developing a variety of highly secure voting systems, which have been labeled *end-to-end* (e2e) voting systems. Although the security properties of the systems were well understood, it was unclear if voters would actually be able to vote using these kinds of high-security systems, because of their novelty.

Three e2e systems were selected for evaluation based on their prominence in the published literature and the fact that they represented three of the major approaches to security that were being investigated by the computer science community. One system, called Helios (Adida, 2008), is a Web-based system that, to enhance security, relies on users performing additional authentication and encryption steps after completing their ballot but before casting it. The second system, Prêt à Voter (Ryan, Bismark, Heather, Schneider, & Xia, 2009), is a paper-based system that randomizes the presentation order of candidates and physically separates the candidate names and bubble selection during the last step of voting. The final system, a version of Scantegrity II (Chaum et al., 2009), utilizes traditional bubble ballots with the addition of embedded security codes that are revealed by special pens when the voter fills in the bubble.

All three systems have the unique feature of allowing voters to verify that their vote has been cast as intended once the polls close, a feature that is not available in any voting systems currently deployed in the United States.

8A) Define the Purpose of the Test

We performed a summative test on the three voting systems because the primary objective was to establish usability benchmarks for these kinds of high-security voting systems. Previous research (Byrne, Greene, & Everett, 2007; Everett et al., 2008) had established usability benchmarks for a wide variety of currently deployed voting systems, including paper ballots, punch cards, electronic voting machines (DREs), and mechanical voting machines. Other research (e.g., Campbell, Tossell, Byrne, & Kortum, 2014) had established benchmarks for undeployed, research-based systems such as mobile phone voting.

Establishing benchmarks for these new high-security voting systems would help researchers understand where these new e2e systems fall in terms of usability in the greater universe of voting systems. This is critically important, as the U.S. presidential election in 2000 clearly demonstrated that usability issues can change election results.

Even though the primary goal was to establish usability benchmarks, we also wanted to gain a greater understanding of the kinds of errors that voters were making (or might make) while using these systems, to provide additional design guidance for future e2e voting systems (even though we couldn't fix any of the errors on these systems, as they weren't developed by our laboratory). Although usability metrics would be sufficient to benchmark the systems, such metrics would not provide additional information to designers about the kinds of issues that might be driving the observed usability characteristics (either high or low).

Toward that end, we wanted to collect information about the usability deficiencies and strengths of the systems, which would enable us to make recommendations for future design efforts. Because we wanted to be able to statistically compare the usability metrics across the three systems with a minimum number of participants, we specified a within-subjects design whereby every user would use each of the three systems in a counterbalanced fashion.

8B) Define the Users

Defining users for this kind of voting study is simple. Federal law describes the exact characteristics of people who are eligible to vote, and we followed that guidance. In the case of U.S. elections, this means that recruited users must be U.S. citizens 18 years old or older and registered to vote in the jurisdiction in which they intend to cast their vote (U.S. Election Assistance Commission, 2014).

This definition means that the user population for this test is essentially everyone, and that users of all ages, genders, socioeconomic levels, and race would be eligible. It is important to note that we would not be generating any requirements regarding the users' skills, knowledge, or abilities in any way. Recall from section 3C that these kinds of broad definitions of users can prove challenging during the assessment, given that variability among users will be high.

8C) Perform a Preliminary Assessment I: Heuristic Assessment

One of the major tasks that had to be completed before any assessment could be undertaken was the construction of each voting systems. Unlike a commercially available product, or a product that has been developed or is being developed in house, no complete turnkey physical versions of these voting systems were available for purchase or use. This meant that we had to build each system (or portions of the systems) before we could begin. We did this by performing detailed literature reviews for each of the systems and talking to their principal designers to create a set of specifications that would allow us to construct reasonable facsimiles of each system in the form envisioned by their designers.

In the case of Helios, the designers provided access to a production platform that would enable us to perform the assessments with minimal effort, mostly in the construction of supporting materials, such as voter check-in, because we could utilize standard laboratory computers for delivery of the interface.

Prêt à Voter required us to construct perforated paper ballots and a variety of scanning and shredding mechanisms to support the procedures specified by their designers. Scantegrity II posed the greatest challenge because it was necessary to create ballots on which a portion of their information was printed in invisible ink. Using special decoder pens, this invisible ink would be rendered visible when voters took specific actions on the ballot. As with Prêt à Voter, for Scantegrity II, multiple processing stations that included scanners and ballot boxes had to be included in the setup of the physical system.

Needless to say, the fidelity of these systems to their designer's intent was of critical importance. Small changes in physical or interface design could have potentially large impacts on voters' ability to use the systems successfully.

Only once the facsimile systems had been constructed and were fully operational could the rest of the assessment process begin, starting with the heuristic assessment. A separate heuristic assessment for each of the systems was conducted. Although comparisons among the usability failings of each of the voting systems would eventually be made, a heuristic assessment is not designed to specify which system is better but, rather, only defines potential issues for a given system.

We utilized the nine general-purpose heuristics identified in Table 3 and added a voting-system-specific heuristic that specified that "the system should engender confidence and trust" because of the importance of this characteristic in voting systems. The output of the entire heuristic assessment for each of the three systems cannot be included here because of space constraints, but some of the outputs for the heuristic assessment (across all three systems) are shown in Table 12.

8D) Define the Tasks

For this test, we defined two top-level tasks for each system:

Task 1: Cast a vote for the candidates listed on this piece of paper.
Task 2: Once the vote has been cast, verify that the vote was counted as cast.

Voting systems are unique in that they have only a single purpose: to afford a voter the ability to cast a vote. High-security e2e systems have the additional feature of allowing a voter to verify that his or her vote has been cast, so this feature was added as a task.

Because Task 1 is too cumbersome to use for subsequent activities, such as timing definition and the cognitive walk-through, we broke it into a number of subtasks. It's important to note that the subtasks were not given to users, only the top-level task; the subtasks were used solely to facilitate the cognitive walk-through and administration of the test. Given that the systems differed in their operation, subtasks for a single system are shown for Task 1 on the Prêt à Voter system. Subtasks of the other systems were similar to the ones shown here. A pictorial representation of the subtasks is shown in Figure 16.

Task 1: Cast a vote for the candidates listed on this piece of paper.
Subtask A: Open the sealed envelope with multiple ballot cards and read instructions.
Subtask B: Mark the ballot cards with their selections.
Subtask C: Detach the candidate list from the marked selections for each ballot card.
Subtask D: Shred the candidate list.
Subtask E: Feed the voting cards into the scanner.
Subtask F: Place the voting cards into the ballot box.
Subtask G: Take the voting receipt from the printer for later verification.

Table 12: Selected Outputs From the Heuristic Assessment Across All Three High-Security Voting Systems

Heuristic	Usability Deficiency that Violates the Heuristic
Employ simple and natural dialog.	• Instructions for the system are lengthy and are written in legalese, rather than easily understandable text.
Speak the user's language.	• The system uses extensive jargon that is specific to voting and to each system (e.g., "ballot token," "cryptographic key").
The interface should have consistency.	• The location of the "next" button changes with each subsequent screen.
Sufficient feedback should be provided.	• It is unclear what voters should do with the voting receipt once it is given to them. • It is unclear when the user has finished voting, and this may cause voters to go through the process repeatedly.
Minimize the user's memory load.	• 27 different codes must be captured by the user for use in a later step, and nothing to support this is provided.
Sequence of events and operation should be evident.	• Multiple options for user action are presented on the first screen, and it is unclear which of these actions is the correct one to take to start voting.
Precise and constructive error messages should be used.	• If an ballot receipt number is mistyped, the error message simply states "not found," with no further explanation, which might lead the user to think the vote wasn't cast.
System should engender confidence and trust.	• System message says that the voter's vote was "received but not yet recorded," decreasing voter's confidence in what the system is actually doing.

8E) Perform a Preliminary Assessment II: Cognitive Walk-Through

With the task definitions complete, we performed a cognitive walk-through for each of the subtasks using Spencer's (2000) streamlined cognitive walk-through method. This was repeated for all three voting systems.

Let's examine the cognitive walk-through output for subtask C (which was to detach the candidate list from the marked selections for each valid card):

1) *Will users know what to do at this step?*
   a) The instruction sheet clearly notes that "after marking all of your selections, detach the candidates list" (left side of cards), but at this point in the voting process, the user may have discarded the voting instructions, or may not have read the instructions at the beginning of the process because of their length.
   b) Each of the cards has a noticeable perforation down the center, which could serve as a cue to tear the card, provided the user reads the instructions.

c) The instruction to detach the candidates list from the marked selections is presented only on the instructions page, and there is no indication on each of the ballot cards that this step must take place.
d) If the user takes the undetached card to the next step (the scanner), it will not fit in the slot. Only half of the card will fit in the scanner.

2) *If the user does the right thing, will he or she know that he/she did the right thing?*
   a) The user will be left with two halves of the original card, indicating that he or she has successfully separated the card.
   b) If the user takes the detached card to the next step (the scanner), it will fit in the slot appropriately.

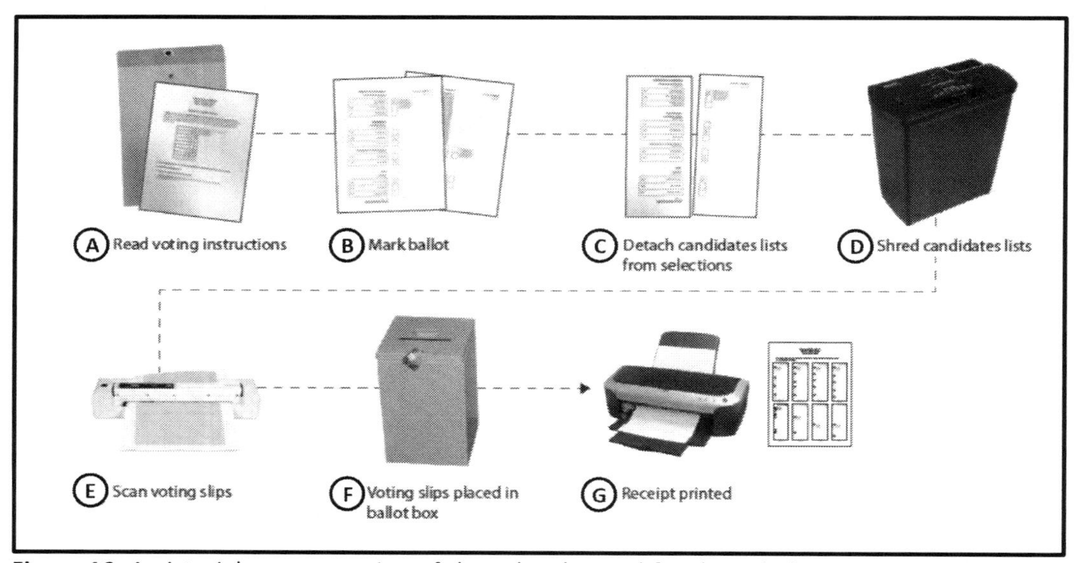

Figure 16: A pictorial representation of the subtasks used for the Prêt à Voter system (modified from Acemyan, Kortum, Byrne, and Wallach, 2015, with permission).

As can be seen from the cognitive walk-through for this specific task, it seems likely that users may have significant difficulty determining what to do with their ballot at this particular step in the process. Although instructions have been provided, they are not provided in a way that would make them easily accessible at the time this step is to take place.

The perforation might serve as an external clue, but if the user did not initially read the instructions, there is no indication that tearing the ballot in half is the next logical step. This is particularly true given the fact that tearing a ballot in half is unique to this system, so even experienced voters would not have a mental model that would suggest this is a valid action. Further, destructive actions in user interfaces can be a problem for many users. If users do make the separation, it appears that that there will be ample evidence that they will know they did the right thing, as they will be left holding two separate pieces of paper, and the physical form of the two pieces allow them to be used in the next step of the process.

8F) Create the Test Plan

The test plan for the voting system included all of the necessary elements required to conduct the usability assessment. Each of these elements is detailed next.

8Fi) Define the Metrics

As in the example of the formative assessment, we used the three metrics of effectiveness, efficiency, and satisfaction, as defined by ISO 9241–11, for both the vote-casting task and the vote verification task. For the vote-casting task, effectiveness was determined by three factors: (a) whether a user was able to cast a ballot successfully, (b) the number of races in which a voter failed to vote as was told to do so using the provided slate (per-race errors), and (c) whether the completed ballot had any errors at all.

Efficiency was determined by the overall time it took a voter to cast a vote, and satisfaction was measured using the System Usability Scale (see section 4Aii). For the vote verification task, effectiveness was determined by whether the user was able to successfully verify that the ballot had been cast. Efficiency was determined by the overall time it took the voter to verify the ballot, and satisfaction was again determined using the SUS.

We did not use the ASQ (section 4A) to measure satisfaction because we reasoned that, given that previous studies had used the SUS alone, it was important for us to maintain the ability to do comparisons with these previously published studies. We would have used more of the secondary metrics of usability (learnability, workload, likelihood to recommend), but because the study was already lengthy, we decided that users were already being overtaxed.

8Fii) Define the Testing Environment

The test was performed locally in a multiroom laboratory facility. Each of the three voting systems was placed in a separate room to prevent users from becoming confused about what equipment and materials were available for a given test. The rooms were not instrumented and did not have one-way mirrors or audio/video recording equipment. The experimenter sat unobtrusively in a corner of the room with a laptop computer to take experimental notes and a stopwatch to record time.

We would have liked to take telemetry from the computer-based voting system (especially for timing), but because we did not have access to the code running on the demonstration server, this was not possible. Figure 17 shows the testing environments of the voting systems.

8Fiii) Define the Timing Parameters

Timing was performed manually using a stopwatch. With efficiency for Task 1 defined as the overall time it took a voter to cast her ballot, timing was started after the experimenter told the participant that she could begin. Timing stopped when the user indicated that she had cast her vote or if she indicated that she wanted to give up. Similar timing landmarks were used for vote verification. These timing landmarks were chosen because they captured all of the activities associated with the actual voting process, and they were clearly identifiable by the experimenter.

Subtask timing landmarks were defined using a single point that marked the beginning of the next task. This means that there was no separate start and stop landmark for each subtask. Rather, the starting landmark for the next task served as the ending landmark for the previous task. These are shown in Table 13.

Figure 17: The testing facility used for the testing of Prêt à Voter. The other two systems were tested in similar rooms located next to this room. Note that the laboratory is not instrumented and does not have a mirror for observation. **(A)** Shows where users marked their ballot, and **(B)** shows the laptop computer where the observer sat and collected data, in the same room (*image copyright by Philip Kortum*).

8Fiv) Define the Testing Materials

8Fiv-1) Product being tested. As described in section 8C, it was necessary for us to construct testable versions of each of the voting systems, as these systems were considered research based and were not available for purchase. Recall that in the case of Helios, a Web-based voting method, we were able to use the company's demonstration system. The other two systems required highly specialized paper ballots, including design features such as perforations and invisible ink, which needed to be printed.

The configuration of the ballot, which was composed of 27 races/propositions, was defined by previous research in the field so direct comparisons in effectiveness and efficiency could be made (Byrne, Greene, & Everett, 2007; Everett et al., 2008). The systems also required ballot boxes, scanning equipment, and paper shredders.

8Fiv-2) Support material and services. Several of the voting systems required that we generate fake voting credentials so users could "vote." One of the systems required an active e-mail address, and we provided Gmail accounts to users so their voting records and credentials would appear in the computerized voting system, just as in a deployed system. None of the systems had training material aside from the instructions provided on the ballot, and no outside help utilities or services were made available to the users. Standard voting booths were used to simulate a common polling station setup.

8Fiv-3) Test materials. We used IRB-approved consent forms and debrief forms to comply with federal guidelines on research involving human subjects. We used the SUS to measure satisfaction and collected demographic data, including age, gender, computer experience, and the kinds of technologies users had used to vote with in the past. A form that listed the way we wanted the voters to vote in each race was developed, and we also had a questionnaire with 49 questions, which probed users' perceptions of voting security and the kinds of technologies on which they preferred to vote. Pens, pencils, and notepads were available for the participants to use at any time.

Table 13: Timing Landmarks for the Vote-Casting Task for the Prêt à Voter System

Subtask	Landmark to *Start* Timing	Landmark to End Timing
Subtask A: Open the sealed envelope with multiple ballot cards and instructions.	When the experimenter tells the user to begin	
Subtask B: Mark the ballot cards with your selections.	When the user makes the first mark on one of the ballot cards	
Subtask C: Detach the candidate list from the marked selections for each valid card.	When the user begins to tear the ballot	
Subtask D: Shred the candidate list.	When the user leaves the ballot marking station and makes the first step towards the ballot shredding station	
Subtask E: Feed the voting cards into the scanner.	When the user takes the first step towards the vote scanning station away from the vote shredding station	
Subtask F: Place the voting cards into the ballot box.	When the user steps in front of the ballot box	
Subtask G: Take the voting receipt for later verification.	When the user touches the vote receipt	When the user tells the experimenter that he/she has cast the ballot or indicates that he/she give up

Data were collected on a standard laptop computer using a customized spreadsheet that was built before the test began. No automated or semiautomated data collection tools were used. A standard handheld stopwatch was used to collect timing measures.

8Fiv-4) Procedures. Because we were testing three systems, total time for the test was set at 3 hours, with short breaks of 5 to 10 minutes between the use of each system. One hour was allotted between participants to reset the systems, make the laboratory "like new," and download any data.

When participants arrived at the laboratory, we followed our written procedure for administering the informed consent and instructions to the users on how the assessment would proceed. This included specialized instructions that described the features of the systems that would enable voters to verify their vote, as no current voting systems allow this function. The instructions were specific on the interaction between the experimenter and the participant. The participants were told they were to cast a vote with the current system, that the experimenter could provide no help, and that if they encountered problems, they should proceed as they best saw fit to complete the task. No other instructions were given to the participants.

A procedure for equipment failure was devised that indicated that any equipment failure during the test would result in immediate termination of the test and the exclusion of the resulting data from the final analysis. If participants didn't show up, they were telephoned and contacted via e-mail to reschedule their session, if possible, and the experimenter waited for the next scheduled participant.

Because the systems were completely self-contained, no dangerous situations were anticipated. However, if an emergency intervention was required, the experimenter who was in the room was instructed to disable the power to the systems to prevent injury. If such an emergency intervention was required, the participant was to be excused (with pay), and the data from that session would be excluded from the final analysis.

When the user finished casting a vote (or indicated as such to the experimenter), he or she was taken to another room to fill out the post-test questionnaire and be debriefed about the experiment. The debrief was an IRB-approved written document, which participants could read and then ask any questions they might have. It described the purpose of the test and how the person's participation aided in the furtherance of that goal in the study. The debrief form also provided contact information in case participants wanted a copy of the final results. Once the users signified that they were satisfied with the debrief, we followed our written procedure to compensate them (using cash envelopes) and obtained written acknowledgment that we had done so.

Because of the configuration of the facility, no test observers were allowed during the usability sessions. The experimenter moderated the session alone, so no team debrief strategies were necessary at the end of the test.

8G) Recruit the Users

We recruited 37 participants from the local area by placing an advertisement in an online advertisement forum (Craigslist). The ad contained a phone number for a line dedicated for the participants to call, which was connected to an answering machine so they could leave a message should they call when the laboratory was closed. We purposely did not use the personal cell phone number of the experimenter for privacy and security reasons. The advertisements stipulated that respondents had to be U.S. citizens and 18 years old or older to vote. No other requirements were specified, and we relied on the random nature of respondents to create a diversified user population for the test.

In the end, the actual demographic distribution of users was broad, with an age range of 21 to 64, a 60/40 gender mix, and a reasonable distribution of race (38% African-American, 27% Caucasian, 22% Hispanic, and 13% other ethnicities). The ease with which we obtained this diverse population can be attributed partly to the fact that participants were recruited from Houston, Texas, one of the most diverse cities in the country (Capps, Fix, & Nwosu, 2015).

8H) Run the Test

Once the systems had been fully set up with all of their attendant testing material and completed procedures, we were ready to begin testing the systems with representative users.

8Hi) Pilot

We started the pilot test by running several undergraduate students (who were 18 and thus technically fit the required demographic), as their participation did not entail any cost. These pilot participants ensured that we had not forgotten anything significant with regard to required material or procedures. When the student participants showed no major omissions, we recruited three additional participants from the general population using the online advertising mechanism. This enabled us to practice recruiting and provided a more diverse demographic to help validate the testing protocols before we began in earnest.

The pilot testing demonstrated the need for some minor modifications, including where we placed material and instructions on how we would move participants between voting systems after they had finished a test. As recommended in section 5B, the data from these pilot users were discarded before the full test was run. At the end of the pilot testing, we were highly confident that we had all the necessary pieces in place to begin full-scale testing.

8Hii) Full Test

After the modifications were made to the protocols following the pilot test, we were ready to begin running the full summative assessment. Recruited participants were scheduled into the laboratory at the rate of two per day, given the length of the test. As participants arrived, they were greeted, brought to the lab, and handed the instruction set. They were randomly assigned to one of the three voting systems and, upon receiving instruction from the experimenter, began the task of voting on that system. Upon completion of the voting task, the participants were asked to verify that their vote had been cast as intended. After they voted on one system, we administered the questionnaire. Once that was completed, the participants repeated the sequence for the remaining two voting systems. At the end of the session, we debriefed, paid, and dismissed them from the study.

Not all scheduled participants arrived at the designated time, and some did not show up at all. For those participants who arrived late, we gauged whether the buffer time between tests was sufficient to allow them to begin the test. It there was sufficient time, we commenced with testing; if not, we rescheduled the participant. No-shows were contacted at least twice to attempt a reschedule and removed from the participant list if multiple attempts to contact them were unsuccessful, in which case an additional participant was recruited. Because of the number of reschedule sessions, the full study took nearly 2 months to complete.

8I) Report the Results

The output from these usability assessments had two separate and distinct audiences, making it necessary to create reports customized for each. Because the research was sponsored by a U.S. federal agency (National Institute of Standards and Technology), data needed to be presented to them in a specific format. The other audience was made up of academics and practitioners who were involved in developing these kinds of high-security e2e voting systems. The major way in which results are communicated with the latter audience is through peer-reviewed publications, and most such publications have their own format requirements. Even so, both of our reports contained the basic information that is specified in the Common Industry Format report, described in section 5D. This included a description of the test objectives, participant demographics, the context of the test, the experimental design, the usability metrics that were used, a description of the data analysis techniques, and a presentation of the results in textual and graphical form.

The reports contained several additional elements not required by the CIF. The first of these elements was a discussion of how the usability deficiencies observed during the testing might be mitigated in future designs. The second element was a discussion concerning future research that might need to be undertaken to better understand some of the issues encountered during the testing. Although these are not explicitly identified for inclusion in CIF-formatted reports, in our experience, they are reasonably common even in industrial settings, given that project teams are typically interested in more than just the results of usability assessment. An example of the output from this study can be found in the *Journal of Election Technology and Systems* (Acemyan, Kortum, Byrne, & Wallach, 2014, 2015).

In the end, voting researchers from around the world are utilizing the data found in this usability assessment to help them design the next generations of secure voting systems in ways that are usable by voters. As these new systems are developed, it is anticipated that further summative assessments on these newly developed systems will be performed to ensure that progress is being made in the usability of these new voting technologies.

8J) Differences Between the Two Case Studies

After reading these two case studies, you'll note that there are not substantial differences between how summative and formative assessments are conducted. The main difference is that in the formative test, information from the cognitive walk-through and the heuristic evaluations was incorporated into the tested product prior to bringing real users into the laboratory. This is because the goal of a formative assessment is to *improve* the product, whereas the goal of the summative assessment is to simply *measure the usability* of the product.

Also recall that the summative test proceeded smoothly (with the exception of the difficulty in getting users to show up), without any major deviations from the plan. This is generally expected for summative tests, given that the products undergoing summative testing are usually deployed or nearly ready to be deployed. More notable is that the formative assessment also proceeded quite smoothly. Often, the prototype systems that are available for formative testing could result in anomalies during the assessment, and that deviations from the test plan need to be undertaken to keep the test protocol moving. In this case, the care that the Web development firm exercised in developing the site and the rigor with which the development team (including the usability professionals) exercised the site prior to running the pilot test helped to ensure a smooth test.

The other difference that isn't explicitly described in the case studies is that, as noted earlier, summative evaluations are typically performed as a single test. The formative evaluation is usually performed as a series of smaller assessments, with fixes to the product being performed in between each of the individual assessments. In both cases, the standard ISO 9241-11 metrics are collected and detailed notes are taken to identify the kinds of usability failures that were encountered, as well as actions users took to overcome those failures.

8K) Suggested Reading

- "Usability of voter verifiable end-to-end voting systems: Baseline data for Helios, Prêt à Voter, and Scantegrity II," by Claudia Acemyan, Philip Kortum, Michael Byrne, & Dan Wallach. *Journal of Election Technology and Systems, 2*(3), 26–56, 2014.
- "From error to error: Why voters could not cast a ballot and verify their vote with Helios, Prêt à Voter, and Scantegrity II," by Claudia Acemyan, Philip Kortum, Michael Byrne, & Dan Wallach. *Journal of Election Technology and Systems, 3*(2), 1–25, 2015.

CHAPTER 9: SOME PARTING ADVICE

The purpose of this book has been to provide you with a solid foundation for how to perform usability assessments in both the field and the laboratory. Much has been written on the subject, so interested readers should avail themselves of the Suggested Reading sections provided at the ends of the chapters to become more familiar with the nuances of usability assessment. That said, if you follow the template provided in this book, you *will* be able to perform basic usability assessments of simple to moderately complex products, services, and systems with excellent results.

Remember that when you start to perform usability assessments, check your feelings at the laboratory door. Your goal is not to "prove" that the product is usable or not usable; rather, your goal is to perform an unbiased assessment to determine if the product has any usability deficiencies and how those might be rectified.

Get into the laboratory early and perform as many formative tests as you can afford to, given your available time and budget. Early testing enables you to have the greatest impact. Although you should always perform a prelaunch summative evaluation for validation, using that as your only assessment will not give you the gains in usability that you desire.

Don't be intimidated by descriptions of others' amazing usability testing facilities. Recall that in case study 1, we used a state-of-the-art, fully equipped laboratory facility to collect the data, but in case study 2, we used nothing more than a simple office space and a human observer in the room. In both cases, we were able to collect high-quality usability data that positively impacted the final product.

The key to high-quality usability testing is rigor. Careful selection of tasks, meticulous collection of the specified metrics, and consistency in how the assessment is performed – adhere to these three tenets, and you can be assured of gaining useful, insightful information that will allow you to make your products better, safer, and more usable.

APPENDIX A

USABILITY ASSESSMENT CHECKLIST

Prepare for the Assessment

- [ ] Define your purpose
  PAGE 25
- [ ] Define your users
  PAGE 27
- [ ] Perform a preliminary assessment 1: Heuristic assessment
  PAGE 29
- [ ] Define your tasks
  PAGE 32
- [ ] Perform a preliminary assessment 2: Cognitive walk-through
  PAGE 34

Create the test plan

- [ ] Define your metrics
  PAGE 38
  - [ ] ISO 9241-11
    PAGE 38
  - [ ] System Usability Scale (SUS)
    PAGE 41
  - [ ] Learnability and workload
    PAGE 44
- [ ] Define the test environment
  PAGE 46
- [ ] Define the timing parameters
  PAGE 49
- [ ] Define the testing material and equipment
  PAGE 50
  - [ ] Product/service being tested
    PAGE 50
  - [ ] Support material and services
    PAGE 50
  - [ ] Test materials
    PAGE 51
  - [ ] Procedures
    PAGE 51
- [ ] Verify your users
  PAGE 53
- [ ] Obtain approval (IRB)
  PAGE 53

Perform the assessment

- [ ] Recruit your users
  PAGE 56
- [ ] Conduct the pilot test
  PAGE 59
- [ ] Conduct the full test
  PAGE 60
- [ ] Report your results
  PAGE 63
- [ ] Post-launch review
  PAGE 65

REFERENCES

Acemyan, C. Z., Kortum, P., Byrne, M. D., & Wallach, D. S. (2015). From error to error: Why voters could not cast a ballot and verify their vote with Helios, Prêt à Voter, and Scantegrity II. *Journal of Election Technology and Systems, 3*(2), 1–25.

Acemyan, C. Z., Kortum, P., Byrne, M. D., & Wallach, D. (2014). Usability of voter verifiable end-to-end voting systems: Baseline data for Helios, Prêt à Voter, and Scantegrity II. *Journal of Election Technology and Systems, 2*(3), 26–56.

Adebesin, T. F., De Villiers, M. R., & Ssemugabi, S. (2009). Usability testing of e-learning: An approach incorporating co-discovery and think-aloud. In *Proceedings of the 2009 Annual Conference of the Southern African Computer Lecturers' Association* (pp. 6–15). New York, NY: Association for Computing Machinery.

Adida, B. (2008). Helios: Web-based Open-Audit Voting. In *USENIX Security Symposium, 17*, 335–348.

Agresti, A., & Presnell, B. (2002). Misvotes, undervotes and overvotes: The 2000 presidential election in Florida. *Statistical Science*, 436–440.

Albert, W., Tullis, T., & Tedesco, D. (2010). *Beyond the usability lab: Conducting large-scale online user experience studies.* Burlington, MA: Morgan Kaufmann.

Andreasen, M. S., Nielsen, H. V., Schrøder, S. O., & Stage, J. (2007). What happened to remote usability testing?: An empirical study of three methods. In *Proceedings of the SIGCHI Conference on Human Factors in Computing Systems* (pp. 1405–1414). New York, NY: Association for Computing Machinery.

ANSI. (2001). *Common industry format for usability test reports* (ANSI-INCITS 354-2001). New York: American National Standards Institute.

Arroyo, E., Selker, T., & Wei, W. (2006). Usability tool for analysis of web designs using mouse tracks. In *CHI '06 Extended Abstracts on Human Factors in Computing Systems* (pp. 484–489). New York, NY: Association for Computing Machinery.

Bangor, A., Kortum, P. T., & Miller, J. T. (2008). An empirical evaluation of the System Usability Scale. *International Journal of Human–Computer Interaction, 24*, 574–594.

Bangor, A., Kortum, P., & Miller, J. (2009). Determining what individual SUS scores mean: Adding an adjective rating scale. *Journal of Usability Studies, 4*(3), 114–123.

Barón, A., & Green, P. (2006). Safety and usability of speech interfaces for in-vehicle tasks while driving: A brief literature review (No. UMTRI-2006-5). Ann Arbor, MI: University of Michigan Transportation Research Institute.

Berardi-Coletta, B., Buyer, L. S., Dominowski, R. L., & Rellinger, E. R. (1995). Metacognition and problem solving: A process-oriented approach. *Journal of Experimental Psychology: Learning, Memory, and Cognition, 21*(1), 205.

Bevan, N., & Raistrick, S. (2011). ISO 20282: Is a practical standard for the usability of consumer products possible? In *Design, User Experience, and Usability. Theory, Methods, Tools and Practice* (pp. 119–127). Berlin & Heidelberg: Springer.

Bogdanich, W. (2010, July 31). After stroke scans, patients face serious health risks. *The New York Times*, A1.

Boren, T., & Ramey, J. (2000). Thinking aloud: Reconciling theory and practice. *IEEE Transactions on Professional Communication, 43*, 261–278.

Borsci, S., Kurosu, M., Federici, S., & Mele, M. L. (2013). *Computer systems experiences of users with and without disabilities: An evaluation guide for professionals.* Boca Raton, FL: CRC Press.

Bradner, E., & Sauro, J. (2012). Software user experience and likelihood to recommend: Linking UX and NPS. *Proceedings of the Usability Professional Association International Conference.* (pp. 1–7). Bloomingdale, IL: Usability Professionals Association.

Brooke, J. (1996). SUS-A quick and dirty usability scale. In P. W. Jordan, B. Thomas, B. A. Weerdmeester, & I. L. McClelland (Eds.), *Usability evaluation in industry* (pp. 189–194). London: Taylor & Francis

Bustamante, E. A., & Spain, R. D. (2008). Measurement invariance of the NASA TLX. In *Proceedings of the Human Factors and Ergonomics Society 52nd Annual Meeting* (pp. 1522–1526). Santa Monica, CA: Human Factors and Ergonomics Society.

Byrne, M. D., Greene, K. K., & Everett, S. P. (2007). Usability of voting systems: Baseline data for paper, punch cards, and lever machines. In *Proceedings of the SIGCHI Conference on Human Factors in Computing Systems* (pp. 171–180). New York, NY: Association for Computing Machinery.

Campbell, B. A., Tossell, C. C., Byrne, M. D., & Kortum, P. (2014). Toward more usable electronic voting: Testing the usability of a smart phone voting system. *Human Factors, 56*, 973–985.

Capps, R., Fix, M., & Nwosu, C. (2015). *A profile of immigrants in Houston, the nation's most diverse metropolitan area.* Washington, DC: Migration Policy Institute.

Carroll, A. E., Marrero, D. G., & Downs, S. M. (2007). The HealthPia GlucoPack™ Diabetes Phone: A usability study. *Diabetes Technology & Therapeutics, 9*, 158–164.

Casey, S. (1998). *Set phasers on stun and other true tales of design, technology, and human error* (2nd ed.). Santa Barbara, CA: Aegean Publishing.

Casey, S. (2006). *The atomic chef and other true tales of design, technology, and human error.* Santa Barbara, CA: Aegean Publishing.

Change the Equation. (2015). *Does not compute: The high cost of low technology skills in the U.S. – and what we can do about it. Reports on the condition of STEM learning in the U.S.* Retrieved from http://changetheequation.org/sites/default/files/CTE_VitalSigns_TechBrief.pdf

Chapanis, A. (1999). *The Chapanis chronicles: 50 years of human factors research, education and design.* Santa Barbara, CA: Aegean Publishing.

Chapman, C. N., & Milham, R. P. (2006). The personas' new clothes: Methodological and practical arguments against a popular method. In *Proceedings of the Human Factors and Ergonomics Society 50th Annual Meeting* (pp. 634–636). Santa Monica, CA: Human Factors and Ergonomics Society.

Chaum, D., Carback, R. T., Clark, J., Essex, A., Popoveniuc, S., Rivest, R. L., Ryan, P. Y., Shen, E., Sherman, A. T., & Vora, P. L. (2009). Scantegrity II: End-to-end verifiability by voters of optical scan elections through confirmation codes. *IEEE Transactions on Information Forensics and Security, 4,* 611–627.

Chopra, P. (2010). The ultimate guide to A/B testing. *Smashing Magazine.* Retrieved from http://www.smashingmagazine.com/2010/06/24/the-ultimate-guide-to-a-b-testing/

Clemmensen, T., Shi, Q., Kumar, J., Li, H., Sun, X., & Yammiyavar, P. (2007). *Cultural usability tests – How usability tests are not the same all over the world* (pp. 281–290). Berlin & Heidelberg: Springer.

Cooley, R., Mobasher, B., & Srivastava, J. (1997). Web mining: Information and pattern discovery on the World Wide Web. In *Proceedings of the Ninth IEEE International Conference on Tools with Artificial Intelligence* (pp. 558–567). Los Alamitos, CA: IEEE.

Corbie-Smith, G. (1999). The continuing legacy of the Tuskegee Syphilis Study: Considerations for clinical investigation. *American Journal of the Medical Sciences, 317,* 5–8.

Coursaris, C. K., & Kim, D. J. (2011). A meta-analytical review of empirical mobile usability studies. *Journal of Usability Studies, 6*(3), 117–171.

Darnell, M. J. (2007). How do people really interact with TV? Naturalistic observations of digital TV and digital video recorder users. *Computers in Entertainment (CIE), 5*(2), 10.

Desurvire, H. W. (1994). Faster, cheaper!! Are usability inspection methods as effective as empirical testing? In J. Nielsen & R. Mack (Eds.), *Usability inspection methods* (pp. 173–202). New York: John Wiley & Sons.

Donahue, G. M., Weinschenk, S., & Nowicki, J. (1999). *Usability is good business.* Retrieved from http://www.yucentrik.ca/usability.pdf

Duh, H. B. L., Tan, G. C., & Chen, V. H. H. (2006). Usability evaluation for mobile device: A comparison of laboratory and field tests. In *Proceedings of the 8th Conference on Human-Computer Interaction with Mobile Devices and Services* (pp. 181–186). New York, NY: Association for Computing Machinery.

Dumas, J. S., & Redish, J. (1999). *A practical guide to usability testing.* Bristol, UK: Intellect Books.

Erickson, W., Lee, C., & von Schrader, S. (2014). *2012 Disability Status Report: United States.* Ithaca, NY: Cornell University Employment and Disability Institute (EDI).

Ericsson, K. A., & Simon, H. A. (1980). Verbal reports as data. *Psychological Review, 87*(3), 215.

Everett, S. P., Greene, K. K., Byrne, M. D., Wallach, D. S., Derr, K., Sandler, D., & Torous, T. (2008). Electronic voting machines versus traditional methods: Improved preference, similar performance. In *Proceedings of the SIGCHI Conference on Human Factors in Computing Systems* (pp. 883–892). New York, NY: Association for Computing Machinery.

Faulkner, L. (2003). Beyond the five-user assumption: Benefits of increased sample sizes in usability testing. *Behavior Research Methods, Instruments, & Computers, 35,* 379–383.

Gao, M. & Kortum, P. (2015). The relationship between subjective and objective usability metrics for home healthcare devices. In *Proceedings of the Human Factors and Ergonomics Society 59th Annual Meeting* (pp. 1001–1005). Santa Monica, CA: Human Factors and Ergonomics Society.

Gilb, T. (1988) *Principles of software engineering management.* Reading, MA: Addison Wesley.

Goodman, E., Kuniavsky, M., & Moed, A. (2012). *Observing the user experience: A practitioner's guide to user research.* Burlington, MA: Morgan Kaufmann.

Grossman, T., Fitzmaurice, G., & Attar, R. (2009). A survey of software learnability: Metrics, methodologies and guidelines. In *Proceedings of the SIGCHI Conference on Human Factors in Computing Systems* (pp. 649–658). New York, NY: Association for Computing Machinery.

Gunsalus, C. K., Bruner, E. M., Burbules, N. C., Dash, L., Finkin, M., Goldberg, J. P., ... & Aronson, D. (2007). The Illinois white paper improving the system for protecting human subjects: Counteracting IRB "mission creep." *Qualitative Inquiry, 13,* 617–649.

Gupta, S., & Zeithaml, V. (2006). Customer metrics and their impact on financial performance. *Marketing Science, 25,* 718–739.

Gwizdka, J., & Spence, I. (2007). Implicit measures of lostness and success in web navigation. *Interacting with Computers, 19*, 357–369.

Han, Y. Y., Carcillo, J. A., Venkataraman, S. T., Clark, R. S., Watson, R. S., Nguyen, T. C., Bayir, H., & Orr, R. A. (2005). Unexpected increased mortality after implementation of a commercially sold computerized physician order entry system. *Pediatrics, 116*, 1506–1512.

Hanson, B. L. (1983). Human factors and behavioral science: A brief history of applied behavioral science at Bell Laboratories. *Bell System Technical Journal, 62*, 1571–1590.

Hargittai, E. (2010). Digital na(t)ives? Variation in internet skills and uses among members of the "net generation". *Sociological Inquiry, 80*, 92–113.

Hart, S. G. (2006). NASA-Task Load Index (NASA-TLX); 20 years later. In *Proceedings of the Human Factors and Ergonomics Society 50th Annual Meeting* (pp. 904–908). Santa Monica, CA: Human Factors and Ergonomics Society.

Hart, S. G., & Staveland, L. E. (1988). Development of NASA-TLX (Task Load Index): Results of empirical and theoretical research. *Advances in Psychology, 52*, 139–183.

Hertzum, M., & Jacobsen, N. E. (2001). The evaluator effect: A chilling fact about usability evaluation methods. *International Journal of Human-Computer Interaction, 13*, 421–443.

Hill, S. G., Iavecchia, H. P., Byers, J. C., Bittner, A. C., Zaklade, A. L., & Christ, R. E. (1992). Comparison of four subjective workload rating scales. *Human Factors, 34*, 429–439.

Holden, R. J., & Karsh, B. T. (2010). The technology acceptance model: Its past and its future in health care. *Journal of Biomedical Informatics, 43*, 159–172.

Hurtado, R. (1992, March 26). *Stocks mixed: Trading error hurts Dow. The New York Times.* Retrieved from http://www.nytimes.com/1992/03/26/business/stocks-mixed-trading-error-hurts-dow.html

Hvannberg, E. T., Law, E. L. C., & Lérusdóttir, M. K. (2007). Heuristic evaluation: Comparing ways of finding and reporting usability problems. *Interacting with Computers, 19*, 225–240.

Hwang, W., & Salvendy, G. (2010). Number of people required for usability evaluation: The 10 ± 2 rule. *Communications of the ACM, 53*, 130–133.

International Organization for Standardization (ISO). (1998). *ISO 9241-11 Ergonomic requirements for office work with visual display terminals (VDTs).* Geneva, Switzerland: Author.

International Organization for Standardization (ISO). (2006). ISO/IEC 25062 *Software engineering – Software product quality requirements and evaluation (SQuaRE) – Common Industry Format (CIF) for usability test reports.* Geneva, Switzerland: Author.

International Organization for Standardization (ISO). (2016). ISO/IEC 25066: (Draft): *Systems and software engineering – Software product quality requirements and evaluation (SQuaRE) – Common Industry Formats (CIF) for other usability test methods upon completion.* Geneva, Switzerland: Author.

International Organization for Standardization (ISO). (2006). ISO 20282-1 *Ease of operation of everyday products-part 1.* Geneva, Switzerland: Author.

International Organization for Standardization (ISO). (2006). ISO/IEC 60601-1-8:2006 *Medical electrical equipment – Part 1-8: General requirements for basic safety and essential performance – Collateral Standard: General requirements, tests and guidance for alarm systems in medical electrical equipment and medical electrical systems.* Geneva, Switzerland: International Electrotechnical Commission.

International Organization for Standardization (ISO). (2007). ISO/IEC 62366:2007, *Medical devices – Application of usability engineering to medical devices.* Geneva, Switzerland: International Electrotechnical Commission.

Jaeger, P. T. (2006). Assessing Section 508 compliance on federal e-government Web sites: A multi-method, user-centered evaluation of accessibility for persons with disabilities. *Government Information Quarterly, 23*, 169–190.

James, J. T. (2013). A new, evidence-based estimate of patient harms associated with hospital care. *Journal of Patient Safety, 9*(3), 122–128.

Jaspers, M. W., Steen, T., van Den Bos, C., & Geenen, M. (2004). The think aloud method: A guide to user interface design. *International Journal of Medical Informatics, 73*), 781–795.

Jeffries, R., & Desurvire, H. (1992). Usability testing vs. heuristic evaluation: Was there a contest? ACM *SIGCHI Bulletin, 24*(4), 39–41.

Kaikkonen, A., Kallio, T., Kelalainen, A., Kankainen, A, & Canker, M. (2005). Usability testing of mobile applications: A comparison between laboratory and field testing. *Journal of Usability Studies, 1*(4-16), 23–28.

Kantner, L., & Rosenbaum, S. (1997, October). Usability studies of WWW sites: Heuristic evaluation vs. laboratory testing. In *Proceedings of the 15th Annual International Conference on Computer Documentation* (pp. 153–160). New York, NY: Association for Computing Machinery.

Kemp, J. A. M., & Van Gelderen, T. (1996). Co-discovery exploration: An informal method for the iterative design of consumer products. In P. W. Jordan, B. Thomas, B. A. Weerdmeester, & I. L. McClelland (Eds.), *Usability evaluation in industry* (pp. 139–146). London: Taylor & Francis.

Kjeldskov, J., & Stage, J. (2004). New techniques for usability evaluation of mobile systems. *International Journal of Human-computer Studies, 60*, 599–620.

Kohn L. T., Corrigan, J. M., & Donaldson M. S. (2000). *To err is human: Building a safer health system.* Washington, DC: National Academy Press.

Kortum, P. T., & Bangor, A. (2013). Usability ratings for everyday products measured with the System Usability Scale. *International Journal of Human-Computer Interaction, 29*(2), 67–76.

Kortum, P., Grier, R., & Sullivan, M. (2009). DSL self-installation: From impossibility to ubiquity. *Interfaces, 80*, 12–14.

Kortum, P., & Motowidlo, S. J. (2006). It takes more than technical knowledge to be an effective human factors engineer. In *Proceedings of the Human Factors and Ergonomics Society 50th Annual Meeting* (pp. 1958–1962). Santa Monica, CA: Human Factors and Ergonomics Society.

Kortum, P., & Peres, S. C. (2015). Evaluation of home health care devices: Remote usability assessment. *JMIR Human Factors, 2*(1), e10.

Kortum, P., & Sorber, M. (2015). Measuring the usability of mobile applications for phones and tablets. *International Journal of Human-Computer Interaction.*

Law, E. L. C., & Hvannberg, E. T. (2004, April). Analysis of combinatorial user effect in international usability tests. In *Proceedings of the SIGCHI Conference on Human Factors in Computing Systems* (pp. 9–16). New York, NY: Association for Computing Machinery.

Lee, D., Moon, J., Kim, Y. J., & Mun, Y. Y. (2015). Antecedents and consequences of mobile phone usability: Linking simplicity and interactivity to satisfaction, trust, and brand loyalty. *Information & Management, 52*, 295–304.

Leveson, N. G., & Turner, C. S. (1993). An investigation of the Therac-25 accidents. *Computer, 26*(7), 18–41.

Lewis, J. R. (1991). Psychometric evaluation of an after-scenario questionnaire for computer usability studies: The ASQ. *ACM SIGCHI Bulletin, 23*(1), 78–81.

Lewis, J. R. (1994). Sample sizes for usability studies: Additional considerations. *Human Factors, 36*, 368–378.

Lewis, J. R. (1995). IBM computer usability satisfaction questionnaires: Psychometric evaluation and instructions for use. *International Journal of Human-Computer Interaction, 7*(1), 57–78.

Lewis, J. R., & Sauro, J. (2009). The factor structure of the system usability scale. In *Human Centered Design* (pp. 94–103). Berlin & Heidelberg: Springer.

Macefield, R. (2009). How to specify the participant group size for usability studies: A practitioner's guide. *Journal of Usability Studies, 5*(1), 34–45.

Mack, R. L., & Nielsen, J. (Eds.). (1994). *Usability inspection methods* (pp. 1–414). New York, NY: John Wiley & Sons.

Malone, T. B., Kirkpatrick, M., Mallory, K., Eike, D., Johnson, J. H., & Walker, R. W. (1980). *Human factors evaluation of control room design and operator performance at Three Mile Island-2* (No. NUREG/CR-1270, Vol. 1). Alexandria, VA: Essex Corp.

Marcus, A. (2005). User interface design's return on investment: Examples and statistics. In R.G. Bias & D. J. Mayhew (Eds.), *Cost justifying usability: An update for the Internet age* (pp. 17–39). Amsterdam: Elsevier.

Masip, L., Oliva, M., & Granollers, T. (2014). Common Industry Format (CIF) report customization for UX heuristic evaluation. In *design, user experience, and usability. Theories, methods, and tools for designing the user experience* (pp. 475–483). Berlin: Springer.

McPherson, I. (2012), *Pedigree technologies blog. Service companies making a buck by not rolling a truck.* Retrieved from http://www.pedigreetechnologies.com/blog/bid/152881/Service-Companies-Making-a-Buck-by-Not-Rolling-a-Truck

Messing, P. (2012, June 24). Unplugged metal detector triggers JFK chaos. *The New York Post.* Retrieved from http://nypost.com/2012/06/24/unplugged-metal-detector-triggers-jfk-chaos-sources/

Miaskiewicz, T., & Kozar, K. A. (2011). Personas and user-centered design: How can personas benefit product design processes? *Design Studies, 32*, 417–430.

Molich, R., & Dumas, J. S. (2008). Comparative usability evaluation (CUE-4). *Behaviour & Information Technology, 27*, 263–281.

Molich, R., Ede, M. R., Kaasgaard, K., & Karyukin, B. (2004). Comparative usability evaluation. *Behaviour & Information Technology, 23*(1), 65–74.

Molich, R., Thomsen, A. D., Karyukina, B., Schmidt, L., Ede, M., van Oel, W., & Arcuri, M. (1999). Comparative evaluation of usability tests. In *CHI '99 Extended Abstracts on Human Factors in Computing Systems* (pp. 83–84). New York, NY: Association for Computing Machinery.

Morrow, D. (2000). *Jury rules against Honeywell, Jeppesen over Cali crash.* Flight Global. Retrieved from http://www.flightglobal.com/news/articles/jury-rules-against-honeywell-jeppesen-over-cali-crash-240775/

National Commission for the Protection of Human Subjects of Biomedical and Behavioral Research. (1978). *The Belmont report: Ethical principles and guidelines for the protection of human subjects of research.* Retrieved from http://www.hhs.gov/ohrp/humansubjects/guidance/belmont.html

National Institute of Standards and Technology (NIST). (n.d.). *Guidelines on how to complete the modified CIF template for Voting System Test Laboratories (VSTLs).* Washington, DC: NIST.

National Institute of Standards and Technology (NIST). (2007). *Usability Performance Benchmarks For the Voluntary Voting System Guidelines (DRAFT).* Washington, DC: Author. Retrieved from http://www.itl.nist.gov/div897/voting/UsabilityBenchmarksWP062607.pdf

National Safety Council. (2013). *Annual estimate of cellphone crashes 2013.* Retrieved from http://www.nsc.org/DistractedDrivingDocuments/Cell-Phone-Estimate-Summary-2013.pdf

Nielsen, J. (1992a). Finding usability problems through heuristic evaluation. In *Proceedings of the SIGCHI Conference on Human Factors in Computing Systems* (pp. 373–380). New York, NY: Association for Computing Machinery.

Nielsen, J. (1992b). Evaluating the thinking aloud technique for use by computer scientists. In H. R. Hartson & D. Hix (Eds), *Advances in human-computer interaction* (p. 82). Norwood, NJ: Ablex.

Nielsen, J. (1993). *Usability engineering.* Boston, MA: Academic Press.

Nielsen, J. (1994a). Heuristic evaluation. In J. Nielsen & R. Mack (Eds.), *Usability inspection methods* (pp. 25–62). New York, NY: John Wiley & Sons.

Nielsen, J. (1994b). Estimating the number of subjects needed for a thinking aloud test. *International Journal of Human-Computer Studies, 41*, 385–397.

Nielsen, J. (2000). Why you only need to test with 5 users. Retrieved from www.nngroup.com/articles/why-you-only-need-to-test-with-5-users/

Nielsen, J. (2001). Success rate: The simplest usability metric. Retrieved from http://www.nngroup.com/articles/success-rate-the-simplest-usability-metric/

Nielsen, J., Clemmensen, T., & Yssing, C. (2002). Getting access to what goes on in people's heads? Reflections on the think aloud technique. In *Proceedings of the Second Nordic Conference on Human-Computer Interaction* (pp. 101–110). New York, NY: Association for Computing Machinery.

Nielsen, J., & Landauer, T. K. (1993). A mathematical model of the finding of usability problems. In *Proceedings of the INTERACT '93 and CHI '93 Conference on Human Factors in Computing Systems* (pp. 206–213). New York, NY: Association for Computing Machinery.

Nielsen, J., & Molich, R. (1990). Heuristic evaluation of user interfaces. In *Proceedings of the SIGCHI Conference on Human Factors in Computing Systems* (pp. 249–256). New York, NY: Association for Computing Machinery.

Olson, P. (2014). Insurers aim to track drivers through smartphones. *Forbes.* Retrieved from http://www.forbes.com/sites/parmyolson/2014/08/05/for-insurers-apps-become-a-window-to-monitor-drivers/

Paolacci, G., Chandler, J., & Ipeirotis, P. G. (2010). Running experiments on Amazon Mechanical Turk. *Judgment and Decision Making, 5*, 411–419.

Paymans, T. F., Lindenberg, J., & Neerincx, M. (2004). Usability trade-offs for adaptive user interfaces: Ease of use and learnability. In *Proceedings of the 9th International Conference on Intelligent User Interfaces* (pp. 301–303). New York, NY: Association for Computing Machinery.

Petrie, H., Hamilton, F., King, N., & Pavan, P. (2006). Remote usability evaluations with disabled people. In *Proceedings of the SIGCHI Conference on Human Factors in Computing Systems* (pp. 1133–1141). New York, NY: Association for Computing Machinery.

Polson, P. G., Lewis, C., Rieman, J., & Wharton, C. (1992). Cognitive walkthroughs: A method for theory-based evaluation of user interfaces. *International Journal of Man-Machine Studies, 36*, 741–773.

Portigal, S. (2008). Persona non grata. *Interactions, 15*(1), 72.

Preece, J., Rogers, Y., Sharp, H., Benyon, D., Holland, S., & Carey, T. (1994). *Human-computer interaction.* Boston: Addison-Wesley Longman Ltd.

Reason, J. (1990). *Human error.* Cambridge, MA: Cambridge University Press.

Redish, J. G., Bias, R. G., Bailey, R., Molich, R., Dumas, J., & Spool, J. M. (2002)). Usability in practice: Formative usability evaluations-evolution and revolution. In *CHI '02 Extended Abstracts on Human Factors in Computing Systems* (pp. 885–890). New York, NY: Association for Computing Machinery.

Reichheld, F. F. (2003). The one number you need to grow. *Harvard Business Review, 81*(12), 46–55.

Rogers, B. (2009a). Unplugged metal detector leads to gun in cell. *Houston Chronicle.* Retrieved from http://www.chron.com/news/houston-texas/article/Unplugged-metal-detector-linked-to-gun-in-cell-1738115.php

Rogers, B. (2009b). 2 juvenile officials expected to leave in detector scandal: A call for "housecleaning" at juvenile probation. *Houston Chronicle.* Retrieved from http://www.chron.com/news/houston-texas/article/2-juvenile-officials-expected-to-leave-in-1745716.php

Rømen, D., & Svanæs, D. (2012). Validating WCAG versions 1.0 and 2.0 through usability testing with disabled users. *Universal Access in the Information Society, 11*, 375–385.

Rubin, J., & Chisnell, D. (2008). *Handbook of usability testing: How to plan, design and conduct effective tests.* New York, NY: John Wiley & Sons.

Rumburg, J. (1998). Good help meta view: Help desks can cut costs and enhance productivity. Is yours meeting the mark? *CIO, 11*(15), 76–79.

Ryan, P. Y., Bismark, D., Heather, J., Schneider, S., & Xia, Z. (2009). Prêt à voter: A voter-verifiable voting system. *IEEE Transactions on Information Forensics and Security, 4*(662–673.

Sanders, M. S., & McCormick, E. J. (1993). *Human factors in engineering and design* (7th ed.). New York, NY; McGraw-Hill.

Sauro, J. (2011a). *A practical guide to the system usability scale: Background, benchmarks & best practices.* Denver, CO: Measuring Usability LLC.

Sauro, J. (2011b) *Usability and net promoter benchmarks for consumer software.* Retrieved from www.measuringusability.com/software-benchmarks.php

Sauro, J. (2012). *How effective are heuristic evaluations?* Retrieved from http://www.measuringu.com/blog/effective-he.php

Sauro, J. (2015). *Are you conducting a heuristic evaluation or an expert review?* Retrieved from http://www.measuringu.com/blog/he-expert.php

Sauro, J., & Lewis, J. R. (2009, April). Correlations among prototypical usability metrics: Evidence for the construct of usability. In *Proceedings of the SIGCHI Conference on Human Factors in Computing Systems* (pp. 1609–1618). New York, NY: Association for Computing Machinery.

Sauro, J., & Lewis, J. R. (2012). *Quantifying the user experience: Practical statistics for user research.* Amsterdam: Elsevier.

Scharff, L. F., & Kortum, P. (2009). When links change: How additions and deletions of single navigation links affect user performance. *Journal of Usability Studies, 5*(1), 8–20.

Schmettow, M. (2012). Sample size in usability studies. *Communications of the ACM, 55*(4), 64–70.

Schumacher, R. M., & Lowry, S. Z. (2010). *Customized Common Industry Format template for electronic health record usability testing* (NISTIR, 7742). Washington, DC: National Institute of Standards and Technology.

Scriven, M. (1967). The methodology of evaluation. In R. E. Stake (Ed.), *AERA Monograph Series on Curriculum Evaluation,* 1. Chicago: Rand McNally

Section 508 of the Rehabilitation Act, 29 U.S.C. § 794d.

Section 255 of the Telecommunications Act, codified at 47 USC § 255

Serafin, C., Wen, C., Paelke, G., & Green, P. (1993). Car phone usability: A human factors laboratory test. In *Proceedings of the Human Factors and Ergonomics Society 37th Annual Meeting* (pp. 220–224). Santa Monica, CA: Human Factors and Ergonomics Society.

Shabtai, A., & Elovici, Y. (2010). Applying behavioral detection on android-based devices. In *Mobile Wireless Middleware, Operating Systems, and Applications* (pp. 235–249). Berlin and Heidelberg: Springer.

Scholtz, J., Wichansky, A., Butler, K., Morse, E., & Laskowski, S. (2002). Quantifying usability: The industry usability reporting project. In *Proceedings of the Human Factors and Ergonomics Society 46th Annual Meeting* (pp. 1930–1934). Santa Monica, CA: Human Factors and Ergonomics Society.

Siau, K. (2003). Evaluating the usability of a group support system using co-discovery. *Journal of Computer Information Systems, 44*(2), 17.

Solomon, S. S., & King, J. G. (1997). Fire truck visibility: Red may not be the most visible color, considering the rate of accident involvement with fire trucks. *Ergonomics in Design, 5*(2), 4–10.

Solomon, Z., Mikulincer, M., & Hobfoll, S. E. (1987). Objective versus subjective measurement of stress and social support: Combat-related reactions. *Journal of Consulting and Clinical Psychology, 55,* 577.

Spencer, R. (2000). The streamlined cognitive walkthrough method, working around social constraints encountered in a software development company. In *Proceedings of the SIGCHI Conference on Human Factors in Computing Systems* (pp. 353–359). New York, NY: Association for Computing Machinery.

Spool, J., & Schroeder, W. (2001). Testing web sites: Five users is nowhere near enough. In *CHI '01 Extended Abstracts on Human Factors in Computing Systems* (pp. 285–286). New York, NY: Association for Computing Machinery.

Surabattula, D., Harvey, C. M., Aghazadeh, F., Rood, J., & Darisipudi, A. (2009). Usability of home cholesterol test kits and how their results impact patients' decisions. *International Journal of Industrial Ergonomics, 39*(1), 167–173.

Tossell, C. C., Kortum, P., Shepard, C. W., Rahmati, A., & Zhong, L. (2012). Getting real: A naturalistic methodology for using smartphones to collect mediated communications. *Advances in Human-Computer Interaction 2012*

U.S. Department of Energy, Federal Energy Regulatory Commission. (1978). *The Con Edison power failure of July 13 and 14, 1977: Final staff report.* Retrieved from http://blackout.gmu.edu/archive/pdf/usdept001_050.pdf

U.S. Department of Health and Human Services. (2009). *Protection of human subjects.* 45 CFR § 46

U.S. Department of Transportation, Federal Aviation Administration. (1994). *Lessons learned: China Airlines flight 140, A300B4-622R, B1816*. Retrieved from http://lessonslearned.faa.gov/ll_main.cfm?TabID=3&LLID=64&LLTypeID=0

U.S. Election Assistance Commission (2014). *14 facts about voting in federal elections*. Retrieved from http://www.eac.gov/assets/1/Documents/Voter'sGuide_508.pdf

U.S. Food and Drug Administration. (2011). *Draft guidance for industry and Food and Drug Administration staff: Applying human factors and usability engineering to optimize medical device design*. Washington, DC: Author.

Van den Haak, M. J., de Jong, M. D., & Schellens, P. J. (2004). Employing think-aloud protocols and constructive interaction to test the usability of online library catalogues: A methodological comparison. *Interacting with Computers, 16*, 1153–1170.

Van der Geest, T. (2006). Conducting usability studies with users who are elderly or have disabilities. *Technical Communication, 53*(1), 23–31.

Virzi, R. A. (1992). Refining the test phase of usability evaluation: How many subjects is enough? *Human Factors, 34*, 457–468.

Virzi, R. A., Sorce, J. F., & Herbert, L. B. (1993). A comparison of three usability evaluation methods: Heuristic, think-aloud, and performance testing. In *Proceedings of the Human Factors and Ergonomics Society 37th Annual Meeting* (pp. 309–313). Santa Monica, CA: Human Factors and Ergonomics Society.

Vredenburg, K., Isensee, S., & Righi, C. (2001). *User-centered design: An integrated approach*. Englewood Cliffs, NJ: Prentice Hall.

Vredenburg, K., Mao, J. Y., Smith, P. W., & Carey, T. (2002). A survey of user-centered design practice. In *Proceedings of the SIGCHI Conference on Human Factors in Computing Systems* (pp. 471–478). New York, NY: Association for Computing Machinery.

Waters, S., Carswell, M., Stephens, E., & Selwit, A. S. (2001). Usability testing. *Ergonomics in Design, 9*(2), 15–20.

Wharton, C., Rieman, J., Lewis, C., & Polson, P. (1994, June). The cognitive walkthrough method: A practitioner's guide. In J. Nielsen & R. L. Mack (Eds.), *Usability inspection methods* (pp. 105–140). New York, NY: John Wiley & Sons.

Wiener, E. L. (1985). *Cockpit automation: In need of a philosophy* (SAE Technical Paper No. 851956). Warrendale, PA: SAE International.

Wierwille, W. W., & Connor, S. A. (1983). Evaluation of 20 workload measures using a psychomotor task in a moving-base aircraft simulator. *Human Factors, 25*, 1–16.

Wiklund, M. E., Kendler, J., & Strochlic, A. Y. (2010). *Usability testing of medical devices*. Boca Raton, FL: CRC Press.

Woolrych, A., & Cockton, G. (2001). Why and when five test users aren't enough. In *Proceedings of IHM-HCI 2001 Conference* (Vol. 2, pp. 105–108). Toulouse, France: Cépadèus.

Zhang, D., & Adipat, B. (2005). Challenges, methodologies, and issues in the usability testing of mobile applications. *International Journal of Human-Computer Interaction, 18*, 293–308.

Zimbardo, P. G. (1973). On the ethics of intervention in human psychological research: With special reference to the Stanford prison experiment. *Cognition, 2*, 243–256.

INDEX

ABOUT THE AUTHOR

Philip Kortum is a faculty member in the Department of Psychology at Rice University in Houston, Texas. His primary interests are in the research and development of highly usable systems in the voting and mobile computing domains and in the characterization of measures of usability and usable systems. Prior to joining Rice, he worked for more than 15 years in the defense and telecommunications industries, where he researched and helped to field award-winning user-centered systems. He is the author of more than 80 peer-reviewed papers and conference proceedings articles and holds 49 U.S. patents. Kortum received his MS from Northeastern University and his PhD from the University of Texas at Austin.